D0952846

HARD GREEN

HARD GREEN

◆

Saving the Environment from the Environmentalists

A CONSERVATIVE MANIFESTO

PETER HUBER

BASIC
BOOKS

A Member of the Perseus Books Group

Designed by Rachel Hegarty

Library of Congress Cataloging-in-Publication Data
Huber, Peter W. (Peter William), 1952-
    Hard green: saving the environment from the environmentalists: a conservative manifesto/Peter Huber.
      p.   cm.
    Includes bibliographical references and index.
    ISBN 0-465-03112-9
    1. Environmentalism. 2. Environmental protection. I. Title.
GE195 .H83 1999
363.7—dc21   99-04225

99 00 01 02 03 / 10 9 8 7 6 5 4 3 2 1

For Sophie, Mike, and Steve

[T]here are no words that can tell the hidden spirit of the wilderness, that can reveal its mystery, its melancholy, and its charm. There is delight in the hardy life of the open, in long rides rifle in hand, in the thrill of the fight with dangerous game. Apart from this, yet mingled with it, is the strong attraction of the silent places, of the large tropic moons, and the splendor of the new stars; where the wanderer sees the awful glory of sunrise and sunset in the wide waste spaces of the earth, unworn of man, and changed only by the slow change of the ages through time everlasting.

Theodore Roosevelt (1909)

[O]nce enough sand is poured to form a single sandpile, and once there is physical contact between all the grains of sand, each new grain sends 'force echoes' of its impact cascading—however faintly—down through the pile, in effect communicating its impact to the rest of the sandpile, causing some grains to shift in position and in the process shifting or reconfiguring the entire sandpile. . . . Is our species now on the verge of a kind of midlife crisis? . . . Are we instead on the verge of a new era of generativity in civilization, one in which we will focus on the future of all generations to come? . . . 'Avalanches' of change . . . are certain to occur and persist if we keep making this sandpile steeper and larger; moreover, the combination of several significant changes occurring almost simultaneously increases the risk of catastrophe significantly.

Al Gore (1992)

CONTENTS

THE ROUGH RIDER
AND THE WONK

As a political movement, environmentalism was invented by a conservative Republican. He loved wild animals. He particularly loved to shoot them.

In the spring of 1908, with time running out on his second term, Theodore Roosevelt convened in the East Room of the White House a hugely successful conference on conservation. The report that emerged, T. R. would declare, was "one of the most fundamentally important documents ever laid before the American people." He promptly convened a hemispheric conservation conference and was working on a global one when he left office in March 1909.

For all his hunting—or was it because of it?—T. R. loved "silent places, unworn of man." As president he resolved to conserve them. Congress had proclaimed Yellowstone a national park in 1872. Yosemite, Sequoia, and General Grant National Parks were established in 1890. The first U.S. forest reserve, a forerunner of the national forests, was proclaimed in the area around Yellowstone in 1891. Presidents Harrison, Cleveland, and McKinley had transferred some 50 million acres of timberland into the reserve system. Roosevelt's distinction was to give conservation its name and, more important, to transform it into an enduringly popular political force. On the way to adding 150 million acres to the country's forest reserves, he would persuade the great mass of ordinary Americans that conservation was in their own best interests.

Fast forward seventy years. Conservation is still high on the political agenda, but its sights are lower. The Resource Conservation and Recovery Act of 1976 addresses garbage dumps, not forests. The 1970s—the post-Vietnam years, years of malaise, oil embargoes, and limits to growth, of Agent Orange and hostages—are coming to a close. Jimmy Carter is president when the new environmentalism officially supplants the old.

It is a time to think small. The new environmentalist still talks about the great outdoors but is obsessed with invisibles. The snarled gas lines precipitated by the Arab oil embargo of 1973 persuade many that a new age of scarcity is dawning. Markets won't recover; the "Invisible Hand" is palsied and in rapid decline. Love Canal transforms the unseen into a national crisis. In 1978 residents of the area are alerted to alarming concentrations of chemicals in the soil and groundwater. President Carter declares a state of emergency and orders the relocation of 710 families. In the waning days of his administration, a panicked Congress enacts Superfund, a massive new program for the cleanup of chemical dumps.

And with that, environmentalism completes its precipitous descent into the siliconized recesses of the microcosm. Practical lessons learned on the plains of Dakota are replaced by pure theory spun out in the Byzantine corridors and computers of Washington. Who now aspires to host the East Room environmental conference of 2008? Al Gore. Yes, that is what it has come to. Theodore Roosevelt, Rough Rider, hunter, Dakota conservationist, gives way to Al Gore, fusspot curator,[1] technorationalist, Kennedy School wonk. If he were alive today, the author of *Hunting Trips of a Ranchman* would waste no time at all with the author of *Earth in the Balance*. Theodore Roosevelt would hardly recognize the environmentalism of the man who now bids to succeed him.

For all its scientific pretension and lofty rhetoric, for all its earnest, computer-powered models and far-future prophecies, for all its cultural bullying and missionary fervor, the new environmentalism is not green at all. Its effects are the opposite of green. Its policies, however well intentioned, do not conserve the wilderness; they hasten its destruction.

Here is the Hard Green environmental manifesto. Expose the Soft Green fallacy. Reverse Soft Green policy. Rediscover T. R. Reaffirm the conservationist ethic. Save the environment from the environmentalists.

HARD GREEN AND SOFT

Theodore Roosevelt learned his conservation the hard way. After Grover Cleveland defeated the Republicans in 1884, T. R. returned to his Chimney Butte ranch in Dakota with plans to increase his cattle herd fivefold. Armed neighbors came by to complain. As H. W. Brands recounts in his 1997 book, *T. R: The Last Romantic*, "the potential for overstocking the range weighed constantly on the minds of the ranchers of the plains."[2] Washington owned the range, but didn't bother to manage it. Among cattlemen, it was first come, first served.

With fists and guns of his own, Roosevelt faced down his angry neighbors. Then he set about finding a political solution to the problem. He formed and was elected president of The Little Missouri Stockmen's Association and wrote a constitution for it. He would only regret not starting earlier. The Dakota pastures were badly overgrazed by the summer of 1886. Many herds, T. R.'s among them, were destroyed in the dreadfully harsh winter that followed.

Two decades later, President Roosevelt and his chief forester, Gifford Pinchot, would be the first to use the word "conservation" to describe environmental policy. By then, T. R. had come to view the misuse of natural resources as "the fundamental problem which underlies almost every other problem of our national life."

With two world wars and a depression intervening, it would take six decades to complete a federal framework for conservation, the work that T. R. began. While they concerned smoke, sewage, and such, the clean air, clean water, and landfill acts of the 1960s were still animated mainly by aesthetics. As with conservation, the principal objectives were still visible and tangible: clear air, swimmable water, and the proper containment of solid waste dumps. New York isn't Yellowstone, but the clarity of its air is still worth conserving, if only because a city can have a horizon, too, sometimes a beautiful one. The Endangered Species Act, passed unanimously by the Senate in 1973, seemed to be cut from the same old conservationist cloth. Cougars, bears, bison: T. R. would surely have voted to preserve them from extinction as eagerly as he had hunted them when they were still abundant. These laws extended the conservationist philosophy to its logical and practical limit.

But even as they completed the legal framework for traditional conservation, these laws also quietly launched the new era of environmentalism in the microcosm. To begin with, regulating pollution of any kind requires a more elaborate regulatory structure than regulating parks and reserves. In 1970, President Nixon established a new Cabinet-level agency, the Environmental Protection Agency (EPA), to take charge. More significantly, each of the new laws also included something quite new: an open-ended "toxics" provision, a general invitation to monitor the micro-environment for insidious poisons and regulate as needed. And though written with cougars and such mainly in mind, the Endangered Species Act had been written quite broadly enough to protect kangaroo rats, too. It would soon be amended to cover not only hunting but also "harming," which a federal court construed to cover "habitat modification." A mere statutory afterthought in the 1960s, the micro-environment is getting entire acts of its own a decade later. The Toxic Substances Control Act is enacted in 1976. Then Superfund, in 1980.

This culminates a forty-year transformation of the environmental movement. The radioactive aftermath of Hiroshima has taught a first ghastly lesson about insidious environmental poison. There follow popularized accounts of industrial Hiroshimas, fallout without the bomb. Rachel Carson defines the new genre in 1962, with the publication of *The Silent Spring*. Micro-environmentalism is born. Not long after, somewhere between Vietnam and Love Canal, the legal infrastructure of the new environmentalism slips into place. Conservation isn't abandoned. It is just overtaken politically, subsumed into something bigger. Bigger because it concerns the very small.

Fish die, frogs are deformed, breast cancers proliferate, immune systems collapse, sperm counts plummet, learning disabilities multiply. Invisible toxics are implicated every time. To be sure, nothing is proved irrefutably, but the proof is good enough for reporters and politicians. For more than a few scientists, too. Harmful effects have been unambiguously confirmed—at very high exposures—for a growing list of factors: radiation, tobacco, asbestos, some workplace chemicals, and some medical drugs. Why shouldn't other factors and low exposures be harmful too? It is postulated that even a single hit from the wrong photon or molecule may set off a chain of biological devastation by damaging the nuclear protein of a single cell.

Environmental degradation is discerned, now, in mass spectrometers, sensitive enough to detect everything everywhere. Its random impacts are reckoned up by Monte Carlo simulation in computers. The computers can multiply the invisible by the infinite to arrive at any number they please. The models can link any human activity, however small, to any environmental consequence, however large: It is just a matter of tracing out small effects through space and time, down the rivers, up the food chains, and into the roots, the egg shells, or the fatty tissue of the breast. Environmentalism happens everywhere, all the time.

Hard Green was and remains traditional conservation, T. R.'s environmentalism, the kind that happens in places we can see and draw on a map. Yellowstone and Yosemite start here and end there. Bison, eagles, and rivers are only somewhat harder to track. Micro-environmentalism happens everywhere. The microcosm is so populous, the forces of dispersion so inexorable, that in every breath we take we inhale millions of molecules once breathed by Moses. At that level of things, everybody is a polluter, and everything gets polluted. Even though no one can see it.

Theodore Roosevelt had no trouble seeing the things that made him a conservationist. Forests were being leveled, ranges overgrazed, and game was getting scarce. Hunters and hikers, cattlemen, farmers, and bird watchers could easily grasp all this, too. People either saw something beautiful in the original wilderness—most did—or else they didn't. The political choices T. R. was urging were, above all, aesthetic. You preserved Yellowstone for the same reason you might some day climb Everest, because it was there.

Soft Green is the realm of huge populations (molecules, particles) paired with very weak (low-probability) or slow (long time frame) effects. Soft Green is the Green of the invisible, the Green of the highly dispersed or the far future. To believe in Soft Green you must either be a savant or put a great deal of trust in one. To the Soft Green, the model is everything. Only the model can say just where the dioxin came from or how it may affect our cellular protein. Only the model will tell us whether our backyard barbecues (collectively, of course) are going to alter rainfall in Rwanda. Only the model can explain why a relentless pursuit of the invisible—halogenated hydrocarbons, heavy metals, or pesticides—will save birds or cut cancer rates. Theodore Roosevelt trades in his double-barreled shotgun for a spectrometer. The cry of the loon gives way to the hum of the computer.

It always is a computer, because Soft Green models are very complicated. They expand in size and complexity as fast as their digital electronic habitat expands in power and speed. The new environmentalism is spawned in the petri dish, breeds in silicon, and is said to feed on DNA. You can't hike, swim, view, or (least of all) shoot it. It's all rather like modern art, as deconstructed by Tom Wolfe in his classic essay, *The Painted Word*. The stuff on the canvas doesn't really become art at all until it's been explained in *The New York Times*. You read about it first. Only then—if you're a person of sufficient discernment and good taste—do you actually "see" what's there.

Because it involves things so very small, Soft Green requires management that is very large. Hard Greens can maintain reasonably clean lines between private and public space. They will debate how many Winnebagos to accommodate in Yellowstone or how much logging, hunting, fishing, or drilling for oil to tolerate on federal reserves, but the debates are confined by well-demarcated boundaries. Everyone knows where public authority begins and ends. Yellowstone requires management of a place, not a populace. Municipal sewer pipes and factory smokestacks require more, but still, a conventionally manageable more. But the Soft Green models are completely different. They are tended by a new oligarchy, a priesthood of scientists, regulators, and lawyers.

With detectors and computers that claim to count everything, everywhere, micro-environmentalism never has to stop. With the right models in hand, it is easy to conclude that your lightbulb, flush toilet, and hair spray, your washing machine and refrigerator, your compost heap and your access to contraceptives, are all of legitimate interest to the authorities. Nothing is too small, too personal, too close to home to drop beneath the new environmental radar. It is not Yellowstone that has to be fenced but humanity itself. That requires a missionary spirit, a zealous willingness to work door-to-door. It requires propagandists at the EPA, lesson plans in public schools, and sermons from modern pulpits. Children are taught to teach—perhaps even to denounce—their backsliding parents.

HARD FACTS AND SOFT

It is at this point that environmental discourse invariably degenerates into fractious quarrels about underlying facts. One side insists that ethylene dibromide, pseudo-estrogen, or low frequency electromagnetic radiation seriously harms human health. The other side says it doesn't. One side says such "pollution" will hurt birds, frogs, and forests, and has already done so. The other side says it hasn't and won't. One side insists we will imminently exhaust our supplies of food, lumber, oil, gas, or genetic stock. The other side says we won't. Hard Green and Soft aren't really so different—they claim to pursue the same green ends—they just disagree sharply on the key underlying facts.

One might suppose that rigorous science and disciplined economics would settle such disputes. But they won't.

Begin with Soft Green science. In the environmental context we are not talking about pharmaceutical drugs or polluted workplaces, where populations are well defined, exposures often high, and at least some effects strong and quick. Soft Green is not about things we take in a potent pill or inhale daily for ten years in a foundry. We certainly know that some of *those* things can be very bad. But the tough part is how to scale down and out, from there to here, from high exposure to low, from short times to long, from small, highly exposed populations to huge ones that are hardly exposed at all. Infinite scope of exposure and zero effect can multiply out to a lot or to nothing at all. Only a model will say for sure. But the models themselves are unsure.

It is easy enough to propose a model, for heavy metals or synthetic hydrocarbons or radiation and for their supposed impacts on children or peregrine falcons. Feed the data into the best computer at hand, and off the time-and-space machine whirls. It is a fair bet that now and again a model will predict things exactly right. It is a fairer bet that most of the time it won't. Although they clarify little else, the overall statistics confirm that environmental toxins of human origin aren't the main cause of anything much. The more industrialized we become, the longer we live, the healthier we grow. The same can't be said for the environment around us, but for reasons that have little to do with trace toxics. When we pave paradise, it isn't any trace poison in the asphalt that kills the flowers, it's the steam roller. Shopping malls do indeed crush nature, but for visible reasons, reasons that even T. R. would readily have grasped.

For every model that predicts apocalypse now, another predicts apocalypse not. As exposure levels drop, predicted biological effects may even flip from bad to good. A bit of carbon dioxide grows the grass; a lot may flood the prairie. Metals banned from the workplace are added to your vitamin tablet. There's a model—quite a credible one, in fact—that purports to prove that a steady dose of low-level radiation, like the one you get living in a high-altitude locale like Denver, or at some suitable distance from Chernobyl, actually improves your health, by impelling your cells to shape up. No pain, no gain, says the model.

In a classic 1972 essay, nuclear physicist Alvin Weinberg coined the term "trans-science" to describe the study of phenomena too large, diffuse, rare, or long term to be resolved by scientific means. They are epistemologically "scientific," and yet—for strictly practical reasons—unanswerable by science. It would take eight billion mice to perform a statistically significant test of radiation at exposure levels the EPA deems to be "safe." The model used to set that threshold may be just right, or it may be way off; the only certainty is that no billion-mouse experiment is going to happen. Even if it did, countless other critical components of the model would remain unverified.

It's the same with any model for very-low-probability accidents: an earthquake precipitating the collapse of the Hoover Dam, say, leading to the destruction of the Imperial Valley of California. Statistical models can be built, and have been, but their critical, constituent parts cannot be tested. It's the same for all the most far-reaching models of micro-environmentalism: on global warming, and ozone, species extinction, radiation, halogens, and heavy metals, whether the concern is for humans or frogs, redwoods or sandworts. Here and there a few things can be nailed down solidly. But for all the big-picture issues the time frames are too long, the effects too diffuse, the confounding variables too numerous. When you are saying something about climate a hundred years hence, the one thing you cannot do is step outside and measure the temperature your model has predicted.

The softness of Soft Green science is rarely apparent in the summaries that most ordinary people read in the mass media. Those accounts invariably convey a high sense of certitude. Scientists are all but unanimous—on the inevitability of global cooling—in 1975. And almost unanimous again—on the inevitability of global warming—in 1992. The papers present Soft Green trans-science straight enough, but in different editions. Sometimes they're published twenty years apart. Yet the cooling/warming flip is quite typical of the business. Soft Green models are invariably weak, manipulable, and perfectly capable of flipping from up to down, big to little, more to less, with the simple twist of an analytical knob. A nuclear power plant is unstable, like a sandpile, declares the model. Or tweak the viscosity knob to prove that the nuke is stable, like a gob of honey.

The problems get worse. Suspect toxins vastly outnumber modelers. Scenarios of mutation, meltdown, and catastrophic collapse can be imagined in every new technology. Foreboding scarcities can be predicted in every natural resource. The list of things we might reasonably worry about grows faster than new rules can be published in the *Federal Register*. But the axiology of science, its priorities of investigation and research, the criteria for what to study and what not to, are matters of taste, budget, values, politics: everything but science itself. Scientific priorities, Weinberg notes, are themselves trans-scientific. So are all the engineering issues, the practical fixes that regulators prescribe. Science will never tell us just how much scrubber or converter to stick on a tailpipe or smokestack, how much sand and gravel at the end of a sewer pipe, how much plastic and clay around the sides of a dump.

Above the Soft science stands the equally rickety edifice of Soft Green economics. In recent years, the Softs have directed comparably elaborate and complex computer models to deconstruct economic life and predict its future. Soft Green science makes far-flung pronouncements about cause and effect; Soft Green economics makes pronouncements of equal sweep and ambition about supply and demand. Here again, Soft thought reaches far beyond the bounds of testing or falsification.

The first book of Soft Green economic faith is the book of scarcity, of "limits to growth," famine, plague, pestilence, and war. We have seen in our times the resurrection of Thomas Malthus, not in the flesh but in the computer. Again and again the computers schedule us to run out: of food, then oil, then tin, zinc, mercury, or wood. Markets for ordinary goods of every description will fail. When the predictions themselves fail to materialize on schedule, the schedule is always extended, by another decade or five.

More recently, obesity and greed have risen to prominence as the new, post-Malthusian metaphor. The problem isn't that growth will soon end, and catastrophically; it is that growth seems destined to continue forever, across every last rain forest and prairie, mountain and swamp. The fault, we are told, lies mainly with America. A mere 5 percent of the world's population lives in the United States. Yet we voracious Americans consume 28 percent of the world's natural gas; 23 percent of its solid fuels; 20 percent of its hard coal; 23 percent of its crude petroleum; 42 percent of its motor gasoline; 26 percent of its electricity; and 10 to 30 percent of its copper, aluminum, and zinc—more than any other country. We drive more cars more miles than any other nation on Earth. Macro-economic numbers like these indict America as the major environmental vandal of the planet, the Softs declare.

Government-imposed "efficiency" is a big part of the answer, according to the latest chapter of Soft Green economic theory. Thus we enter the new era of Soft "conservation," in which the lakes, trees, mountains, and prairies are to be saved not directly, but rather through the conservation of gasoline, newspaper pulp, aluminum, and glass and the relentless sifting and sorting of trash.

But buried in all Soft Green economic discourse are long chains of conjecture and projection, most of them disconnected from verifiable economic theory or historical experience. Hard economics affirms that consumption does not presage exhaustion; the demand side of a market tells us nothing about the future of the supply side. Hard economics recognizes no correlation at all between consumption and vandalism of any kind. Hard economics affirms no association at all between efficiency and lowered consumption: frugality. Saving fuel and trash does not save the Earth, not on the Hard economic books. The Soft Green facts, figures, and projections that assert the opposite are linked here by nothing more than unverifiable tissues of speculation. Every time the real world catches up with the pronouncements and disproves them, the details are changed, the timetables revised, and the same pronouncements are repeated with even more conviction. There is no day of reckoning in Soft Green economics. The books of account are never totaled or closed. Anything in the black today will be in the red tomorrow, or fifty years hence. Where Soft Green economic projections can be tested at all, they turn out to be wrong. The price of oil is projected to double in the 1980s; instead, it is cut in half.

End to end, across science and economics, the predictions and prescriptions of Soft Green science and Soft Green economics are so volatile one generally cannot trust even the *sign* at the front end of the prediction: positive or negative, up or down, hot or cold, more or less. Yesterday, the planet's destiny was to end in ice, but today the destiny is fire. Yesterday the perceived crisis was mass famine; today the perceived crisis isn't that the population will soon collapse, it's that it will grow without limit and bury itself in its wastes. Yesterday the crisis was that oil prices were going to spiral up; today the crisis is such that we must slap a big tax on oil because prices have spiraled down.

Occam's razor exhorts us to favor theories that explain the most with the least, the single equation that explains both falling apples and falling planets. Modern environmentalism operates on the opposite principle: ever-increasing complexity and opacity. In modern environmentalism, ten thousand equations are invoked to predict fluctuations in a single thermometer. The web of silicon-simulated life exists only as millions of lines of code, endlessly groomed by a tiny club of digirati. And objective though they endeavor to be, the Soft Greens have much to gain by predicting momentous change and much to lose by admitting how speculative their models remain. The modelers often begin with important insights and real data that deserve sober attention. But their big-future scenarios do not increase that knowledge, they reduce it, by embellishing and projecting it far beyond what it will support.

The reams of statistics, the ever-unraveling models, are of little environmental consequence, Hard Greens maintain. Fuels and minerals are not green, molecular traces are not the important enemies of green, and consumption itself has nothing to do with anything. Harping away at such things just distracts and confuses. The only scarcity that matters in the environmental debate is scarcity of green: of wilderness and forest, lake and river, marsh and shore, of places undeveloped by markets and untouched by the hand of man.

Theodore Roosevelt did not finish the job of conservation; it is not a job that can ever be finished. And until quite recently, we have continued to cut down and consume far more of nature than we have conserved and re-grown. People themselves are not starving, but in many places on the planet the Malthusian misery has been redirected instead at forests, water tables, soil, wetlands, fisheries, range lands, rivers, and coral reefs. "It took but a second to cut off Lavoisier's head," one of his countrymen remarked in 1794, when the great chemist was led to the Paris guillotine. "It will take France a century to grow another like it." And how long to grow another rain forest? How long to replenish the seas? Far longer than it is taking to burn them down and fish them clean, if it can ever be done at all.

To believe in Hard Green we merely have to love the outdoors, the unspoiled wilderness, forest, river, and shore. It takes no special expertise, no extraordinary discernment, and not much science. It requires no big model of the far future. All it takes is an eye for natural beauty, a reverence for life, a sense of awe for the boundless grandeur of creation. We don't preserve Niagara for the extra megawatts we might someday squeeze out of it. The far-future economic case for biodiversity is negligible, but the case for conservation is still compelling. We revere life on Earth not for speculative value in the far future but for tangible value in the present. Something in our own nature finds redwoods, cougars, and whales beautiful and worth having. Life is a fascinatingly complex good that requires no further justification. We conserve because it is there and because we find it magnificent—today.

HARD PATH AND SOFT

The prescriptions that emerge from Soft Green theory are as changeable as the weather. Foreboding scarcities are predicted in one resource after the next. New efficiency mandates are cooked up willy nilly. The list of things in need of prescription grows, yet again, much faster than new rules can be published in the *Federal Register*. All in all, the policies that emerge from the snarl of the Soft Green computer have done far more harm than good. The Softs dress their big-future models up as rigorous physics, chemistry, biology, and economics, but what they prescribe remains as irrational as alchemy. It is as wasteful and destructive, too.

Dumps overflow, forests fall, mines are exhausted? This calls for less earth, more fire. Cast out paper diapers, swaddle your babies in cloth. Which you wash in water that has been heated with oil. The sealed juice pack gets dumped too, so the green alchemist demands recyclable glass instead. You stick the bottle in the fridge, which means you burn more electricity. And diesel too, in the truck that picks up the empty.

Running out of diesel? Conserve that too, but somewhere else—by using less fire, more earth. Ground the fuel-gulping plane, subsidize fuel-efficient trains. Which devour a whole lot more land, for all the track. And turn down your furnace, too; a thicker wall will keep you warm. Asbestos is out of favor, but fiber glass hasn't been condemned. Not yet.

Which reminds us of pollution. Stop squandering air and water! Pile on earth and fire. Install more scrubbers, which create mountains of sludge. Use more catalytic converters, made out of platinum, gouged from the soil of South Africa. You'll use more coal and oil, too. All the extra hardware on the smokestack and tailpipe kills efficiency.

Soft Green alchemists love to prescribe not only the mix of ingredients but the size of the mixing bowl. Cars are too small; use trains. Nukes are too big; use stoves. If it's centralized, disperse it. If it's dispersed, centralize it. Each good deed, alas, entails another. The train needs more than a stove to propel it. France uses electricity for its bullet trains. And breeder reactors for its electricity. Plutonium is a by-product.

When they failed to ennoble lead, the alchemists of old got into pharmacy, prescribing elixirs to restore health. The Soft Greens are still at it. Shun the cancerous pesticide; grow organic. Crop yields fall, so you drive more tractors to plow more acres of land. Eschew malignant growth hormones for cattle. You'll use more pasture, get less milk, but you can claim to have transmuted Ben and Jerry's into a health food. Don't irradiate strawberries to stop spoilage; radiation is poison. Just stick the berries in the fridge. Next to the juice.

And search on, ever on, in your quest for the perfect fire. Uranium burns too hot, coal too dusty, smoky and dry, the Hoover Dam is too wet, and if fusion ever happens, you can bet it will be too cold. The only soup that's just right is the one not yet on the table. Passionate outdoorsman William O. Douglas didn't like the energy alternatives of the 1960s. He preferred the energy of the future: nuclear power.

The Soft Green does part company with the alchemist on the subject of gold. Capital is the ultimate poison. There's too much of it in a big central power plant and too little labor. Windmills and solar cells are better: They employ the working man. It's a matter of putting people first. Unless you're putting them in cars. Big cars are safer, but small ones are greener. Burn people, not gas.

Enough already. I have indulged in a tirade for the last page or two only because the Softs are so good at such polemics themselves.* We Hard Greens believe quite as passionately as Softs, and when it comes to polemics, we can give as good as we get. Let's now lower the noise level a bit and try to define, sharply and cleanly, how Softs and Hards differ in the green policies they prescribe. Our differences fall into four distinct categories.

We divide, first, on what to do about the problem of scarcity. Softs discern it everywhere: scarcity of food and oil, wood and aluminum, paper, glass, and space itself, the space needed to contain all our trash. So they prescribe scarcity solutions everywhere, too. Convinced that markets fail left and right, the Softs set about moving the production of ordinary goods—food, transportation, power, and countless others—out of the market and into the public sector.

Hard Greens do not believe that the far future of markets can be discerned at all, but trust the market will somehow keep on supplying anything it can package and trade. There are no limits to humanity's growth, at least none set by the external environment. Yes, we will run out of what we now consume, eventually. Before we do, we will grow, find, or invent other things.

For environmental purposes there is only one looming scarcity that is important, and that is the scarcity of green. And the one plenty government can effectively promote in this sphere is plenty of nothing. Government can, in other words, play a necessary and desirable role in husbanding and expanding the conservative commune—the wilderness, broadly de-

---

*My own tirade set forth above elicited the following response from a reader when I presented it in *Forbes*:

Dear Mr. Green,

You are painfully stupid. People like you don't understand anything until you get diagnosed with cancer and try to pinpoint the cause. You should build yourself an asbestos house next to a nuclear power plant in a town where you can only drive diesel-power vehicles, must use lead paint in your house and must store all of your waste in your own backyard, including your personal sludge. You can burn the waste but must stand very close to the fire so that you absorb all the wonderful smells of burning plastic—and please, do not attempt to capture the heat for any other purpose. Your town must also pay for the full costs of the nuclear power plant and store that waste in your town. Should be a really nice place to live for those like yourself. A 21st Century utopian society.

You are a half-educated moron.

fined—because the objective there is to see to it that in those places nothing much is done at all. Nothing is the one thing that big government is capable of doing quite well, and doing nothing is the paramount objective of conservation. The whole point of conservation, one might say, is to be economically inefficient and unproductive, to retard conventional economic progress, not promote it, and to do so in well-designated places set aside for that specific objective. Even conservatives can believe in government's ability to do *that*. Theodore Roosevelt certainly did.

The second division between Hard and Soft is less sharp, but no less important. It lies at the blurred boundary where "pollution" gets too attenuated for science to fathom or the market to contain, where the thick smoke and raw sewage give way to ephemeral wisps and molecular traces. Soft Greens recognize no such line. They discern market failure—"externality"—everywhere. They believe that only government can set things right and can set it *all* right, every last wisp and trace and molecule of it. They believe in devoting more and more to the pursuit of less and less, forever.

Hards are concerned about pollution and see externalities everywhere, too, but prescribe more market, not less, and only up to a point. Stopping points are as central to the Hard Green pollution agenda as starting points. Hards agree that abating large, serious, well-defined sources of pollution is efficient and green. To that end, we prescribe new forms of property and new markets. But we also assert that as pollutants shrink and disperse, the costs of the pursuit rise all too fast. Returns don't just diminish, they go negative. Pollution abatement is an industry, too, and that industry is quite as able as others to waste resources, foul air, dirty the water, and despoil the landscape, even as it loudly proclaims its dedication to cleanliness. Burn down the house to roast the pig. To our eyes, Softs labor to make the abatement of important pollutants as expensive and inefficient as possible, and they demand profligate waste in abating pollutants of no importance at all. Hard Greens will have none of that. We will fight the Softs on this front not merely because what they do is wasteful, but because it is environmentally perverse, because it makes things less green, all in all, not more so.

Our attitudes toward technology define the third great divide between Hard and Soft. Softs believe that nuclear power, pesticides, and genetically engineered potatoes are brittle, unreliable, unpredictable, and unstable. They favor wood, gas, and organic agriculture: low technology over high, soft technology over hard. Indeed, I owe my Hard/Soft Taxonomy of Green to Amory Lovins, who gained fame in the 1970s promoting "soft" sources of energy like wind and solar over "hard" ones like uranium and coal.

In our view, Softs man the barricades against the technology most likely to improve life on Earth in the here and now, making men richer and the planet greener, both at once. Softs block high technology, which is frugal and efficient, and promote low technology, which is profligate and wasteful. They insist on burning live fuel rather than dead. Hards insist on just the opposite: Hard technology saves the Earth. Softs believe in gathering fuel from the surface and in traveling (if they must) by train. Hards believe in digging deep for their fuel and flying high to get from here to there. We promote the conservation of wilderness and the preservation of species by living in three dimensions, rather than in two, by scraping less from the living surface and living more off the sterile depths and heights.

The fourth and last intellectual chasm between Hard and Soft concerns efficiency and frugality, wealth and poverty. The Softs maintain that efficiency is environmentally frugal. They believe that human privation promotes environmental wealth, even if they rarely dare say so directly. Softs are sure that efficiency alone will do much to save the environment: efficient furnaces and cars, efficient refrigerators and lightbulbs, even efficient trash, trash so efficient it is transformed back into valuable resources by the magic of a Soft Green recycling truck. They believe in trickle-up environmentalism, the notion that what you save in your more efficient refrigerator will become savings at the power plant, which will become savings at the coal mine.

Hard Greens don't believe a word of it. To our eyes, the Softs' efficiency crusades are a useless distraction at best and positively harmful insofar as they divert attention and promote a false illusion of green progress. In their endlessly meddlesome crusades for efficiency, the Softs are peddling the ecological equivalent of *Thin Thighs in Thirty Days*. The energy saved on a more efficient refrigerator trickles all too easily into a larger one, just as the calories saved with a Diet Coke generally trickle into a brownie. Efficiency is not frugality. Forcing efficiency upon consumers does nothing to make them more frugal, except insofar as it makes them poorer. And it is not poverty that makes people green. It is wealth.

Yes, wealth. Get past all the wonkish debates about scarcity, externality, complexity, and efficiency, and it all comes down to something as basic—and politically polarizing—as rich and poor. We affirm that there are two limits to growth, Soft limits and Hard ones. The Soft limits are the limits of human knowledge, the limits of ignorance, superstition, and anti-science. Poverty is the Soft limit to growth. There are Hard limits, too, and they are better ones. The finer side of our own sensuality limits growth too: our sense of what is beautiful, humanity's innate appreciation of what makes life worth living. Wealth limits growth, too.

That is the most fundamental difference of all. To the Softs, wealth is the problem, not the solution. Limit wealth to limit growth to protect the planet. Hards affirm the opposite. Promote wealth to limit growth to protect the planet. It is wealth—not poverty—that limits family size, limits obesity, limits pollution, limits waste and inefficiency, limits personal consumption. It is the rich, not the poor, who pour their wealth into green. The richer we get, the farther the footprint of our wealth extends: to our children, to our neighbors, then our lands, shores, rivers, lakes, and oceans. Wealth solves the problem of scarcity with abundance.

We can't both be right. The Softs say we promote profligate consumption, reckless pollution, and perilous complexity. We say they corrupt ecological science, muddle green objectives, obstruct green technology, and prescribe policies whose actual effect is inimical to the environmental objectives they so earnestly articulate. They say we pursue nothing but private profit. We say they pursue nothing but bureaucratic power. To our eyes, Soft Green is a political movement whose overall impact on the environment is not green at all; it is quite the opposite. In the end, the Soft Green regulator just names his favorite poison and gets on with regulating it. The process is arrayed in all the sumptuary of science, but all the key calls are political. Soft Green ends up as a pursuit of politics by other means.

HARD POLITICS AND SOFT

Nothing wrong with politics, of course; T. R. reveled in them. He approached conservation politics the way he approached everything: self-confident and assertive, with a grand sense of public purpose unprecedented in the presidency before his time. He delighted in using the stroke of a pen to designate new federal tracts as game reserves. He summarily banned hunting on Florida's Pelican Island when he learned that the showy waterfowl there were endangered by hunters going after their plumes.

Now, just how much we want done with the stroke of a presidential pen is a fair subject of political debate. So too is how much we should hunt, log, ranch, or drill on public land. All the choices conservationists make are conventionally political. In 1998, the Clinton administration designated as a national monument a vast stretch of land in Utah, from Bryce Canyon to the Colorado River and from Boulder to the Arizona state line. It was a controversial call: The area includes the Kaiparowits Plateau, where a Dutch-owned concern was slated to begin mining massive coal formations. Roosevelt would have understood the controversy over Kaiparowits Plateau and approved, not just the decision to conserve, but the in-your-face political courage it required.

Conventional political process decides little in the new environmentalism. The toxics clauses inserted as an afterthought in the Clean Air and Water Acts and as the central thought in Superfund are just a stew of demented words. They articulate no standard, set no budget, establish no limits. They wouldn't even have passed constitutional muster in T. R.'s day. The Supreme Court would have cited the "non-delegation doctrine," which, in those days at least, forbade Congress to delegate responsibilities wholesale to the executive branch.

The delegation today goes a lot further than that. Although nominally overseen by the president, political authority for micro-environmental matters is now centered in the new trans-scientific oligarchy. The key calls are still stroke-of-the-pen political, at bottom, but no ordinary observer can see to the bottom. Roosevelt would readily have admitted that his environmentalism was ultimately a matter of political will. The EPA's enormous staff never will.

It is not a good thing to cut politics out of an essentially political process. The end of politics by other means is politics in disrepute. Government itself comes to be perceived—correctly—as the enemy.

Normal politicians certainly know how to waste money, but only so much. It may or may not make sense to mine coal on the Kaiparowits Plateau, but we surely get *some* real benefit from choosing not to. And unlikely as it may sound in these cynical times, the political resolve behind the decision not to may help enrich us, too. Disney would operate Yellowstone just fine, but designating it a national park helped define a nation Americans could call great. Nothing equivalent happens with our micro-environmental resolutions, because there are no resolutions, none ordinary Americans can latch onto; there are just programs, staffs, offices, and legions of lawyers. The only thing ordinary Americans may dimly realize is that somewhere deep in the EPA it has been deemed wise to spend more digging up an industrial park in New Jersey than ever was spent conserving a forest in the Adirondacks.

Similarly, politicians know how to reward friends and punish enemies, but democratic politics as a whole is pretty evenhanded. When the old conservationists took your land, they paid you for it, and the money came from taxes and user fees. That was about as fair as the income tax: not very, but fair enough. In the new environmentalism, most of the taxing occurs off the public books. There is a great deal of creeping, uncompensated expropriation and a freakish rain of ruin on those unlucky enough to discover the wrong rodent, marsh, or buried chemical on their land. Any amount of public environmental good, however small, can entail any private financial burden, however large.

We have likewise lost, somewhere in the Soft Green microcosm, all pragmatic sense of when enough is enough. Driven as it must be through normal political channels, conservation can be pushed only so far. Roosevelt had added 150 million acres to the forest reserves by 1907, when logging companies finally persuaded Congress to curtail the 1891 law that had given him the power to do so. Tied to a politically untouchable agriculture bill, the new measure couldn't be vetoed. With the time running out on his power as "lord of the forests," T. R. issued a decree expanding the reserve system by 16 million "midnight acres." "If Congress differs from me," he declared, "it will have full opportunity in the future to take such position as it may desire." The Clinton administration had to trade political chips for the Kaiparowits Plateau; nobody fears it will soon seize the rest of Utah. Conservation works, politically, because the boundaries are reasonably well defined, because it targets real estate, not molecules.

But most of the Northeast could be placed in regulatory receivership for its countless micro-environmental derelictions. Hikers and hunters just occupy a seat or two at the political table; synthetic estrogens and carbon dioxide have somehow escaped from the coils of politics, and the priesthood can pursue them anywhere. Under Superfund, the best guess is that off-budget spending totals many times direct federal disbursements. But nobody knows for sure. What we spend is determined only by how big a legal staff the EPA can hire.

The "remedial" efforts that emerge end up harming even the intended beneficiaries. Contact with Superfund has become socially poisonous. The very fact that the EPA has arrived shatters property values, repels new industrial investment, and throws a region's entire future into doubt. Communities once eager to see wastes removed now often prefer to see EPA depart instead. Meanwhile, the culprits targeted by such a process invariably end up wondering: Why me? And so they well may. For all the scientific veneer on things, micro-environmental regulation often ends up irrationally vindictive. Normal democratic politics do not spawn vendettas. Priesthoods often do.

The Softs have led us down Friedrich Hayek's "Road to Serfdom," from good intention to gross bureaucracy, from grand principle to raw power. Soft Green environmental regulation has become a mirror image of the problems it is supposed to solve, a repository of spendthrift political declarations, abandoned on the insecure premises of EPA and the courts. It leaks into society cancerous plumes of lawyers, administrators, and consultants, the brokers of ignorance, speculation, and uncertainty. It does not make things greener in the short run, and it will make them less green in the long.

To Hard Green minds, green does not emerge from big computer models or from large government agencies. Green objectives are effectively advanced only by dispersed control, free markets, and traditional ethics, the conservative instruments for managing the problems of scarcity, dispersion, complexity, greed, growth, consumption, fecundity, and human voracity. Soft Green doesn't make the planet greener. Hard Green does.

Populist Green

Politically, the most important principle is that conservationists can be populist and should be. Soft Greens can't and won't. Their mission is exclusionary. Ours is inclusionary. We welcome humankind as an integral and legitimate part of nature's landscape. We espouse Winnebago environmentalism. We do not see man as tapeworm in the bowel of nature. Symbiosis is possible. And when we have to choose, as we sometimes must, we put people first.

Soft Greens will vehemently deny any suggestion that they do otherwise. They have successfully conflated conservation with micro-environmentalism, eagles with snail darters, halogenated hydrocarbons with the mountain peaks of Yosemite. By conflating them, they have come to stand for them all, in the public mind, in political alignment. If you don't take the snail seriously, you can't be serious about Yosemite. Soft Greens have mastered big-tent environmental politics, leaving T. R. conservationists with not much political tent at all.

Conservatives have to win back the conservation faction of environmental politics. This is essential for their political survival. It is equally essential for the cause of conservation.

Green politics are good politics. Environmentalism is hugely popular. This should not surprise conservatives. Despoliation and decay always have a public dimension, whether they destroy wetlands, wolves, or the human spirit. Conservatives have no trouble grasping society's collective concerns about cocaine in private veins, abortion in private wombs, and the river of filth that flows out of Hollywood into our private televisions. Rivers and redwoods have public implications, too. Green politics are popular because ordinary people grasp that much immediately. And because they consider rivers and redwoods beautiful.

The trouble is, green politics are much easier to fit into a Left-wing political shoe. Manatees, Monarch butterflies, or the Shenandoah River aren't easy to privatize. It can be done, but it takes hard work, and meanwhile some measure of collective protection is essential. The Right hesitates; not the Left. Running the whole "environment"—literally, "that which surrounds"—is an opportunity the Left gladly welcomes. The micro-environment is the best part of all. It requires regulatory agents as subtle and cunning, regulations as ubiquitous and insidious, as the targets they must pursue. It can be pervasively bureaucratic, manipulative, and intrusive. For people who like big government and are part of it, this is political ambrosia.

In political reply, the Right has nothing better to offer than a long tradition of creating parks, husbanding wildlife, and venerating natural heritages of every kind. Nothing but the legacy of T. R. Politically speaking, that should be enough. It is the old conservation, not the new, that welcomes the family in the Winnebago. It is the old that dispenses with oligarchy and caters to the common tastes of the common man. Norman Rockwell is better politics than Robert Mapplethorpe.

And in a democracy, populist green is the only green that will endure. The more the Soft Green priests talk about saving humanity, the faster the common man should count his tulips. Like the Left everywhere, Soft Greens want to save people from the top down, by instructing them. They mean to govern well, but (to paraphrase Daniel Webster) they mean to govern. They promise to be good masters, but they mean to be masters.[3] In a democratic society, ordinary people won't stand for it, not for long. And a good thing, too. Experience teaches that when all is said and done, the Left consumes, exhausts, impoverishes, and despoils. Behold the land once called East Germany: Love Canal, border to border, perfected by Communists.

Too-eager collectivists never end up conserving anything; only the reluctant ones do. The old green, of parks and forests and Winnebagos advances the green cause because of the Winnebago. The man in the Winnebago sees something in it for him. He is enlisted in the cause by appeal to his own private sense of what is beautiful.

Theodore Roosevelt once complained that the "nature fakers" didn't know "the heart" of wild things. "Every time Mr. Roosevelt gets near the heart of a wild thing," one of the targets of his scorn responded, "he invariably puts a bullet through it." Roosevelt himself would later try to rationalize his hunting as necessary for food or for scientific research, but it wasn't really any such thing. The truth of the matter is that T. R. took atavistic pleasure in hunting game in the wilderness. He loved nature the way a man first loves a woman: selfishly, because he finds her beautiful and exciting, because he needs her desperately, however little she may need him. And what is wrong with that? Nothing. It gave us a president who so loved to shoot wild animals that he resolved to conserve the vast open spaces in which they live.

The objectives are conservative. All the science that matters, and that can be trusted, is simple, solid, and straightforward: in short, conservative. The economic logic is conservative. The political origins of the movement are conservative. The enormous political opportunity is there, an opportunity for conservatives to wrest from the impoverishing, environmentally destructive clutches of their political opposites.

The Hard Green environmental manifesto is conservative. Expose the Soft Green fallacy. Reverse Soft Green policy. Rediscover T. R. Reaffirm the conservationist ethic. Save the environment from the environmentalists.

PART I

What Is and What Will Be

I

Scarcity:
Malthus on a Chip

---◆---

"Cancerous growths demand food; but, as far as I know, they have never been cured by getting it." So declared Alan Gregg, a respected vice president of the Rockefeller Foundation, in 1955. The cancer he was referring to was humanity itself. Human population, Gregg argued, spreads over the surface of the Earth like a metastasizing tumor. All attempts to increase resources are ultimately useless.

Even in 1955, the thought was hardly new. Thomas Malthus supplied the original script, in a hastily written pamphlet published anonymously in 1798. Population increases geometrically, food supplies only arithmetically. Sooner or later, the growing gap between supply and demand must end in war, famine, and general misery. If there is one thing certain on this Middle Earth, it is that geometric growth cannot continue forever. The laws of mathematics forbid it. Or that, at least, has been—and remains—the most fundamental article of faith in Soft Green environmentalism. "Growth," declares the renowned Stanford biologist Paul Ehrlich, is "the creed of the cancer cell."

Hard Greens love Ehrlich. The first order of business in most Hard Green screeds is to quote Ehrlich's unfortunate 1968 bestseller, *The Population Bomb*.[1] It is always the same quote. "The battle to feed all of humanity is over. In the 1970s and 1980s hundreds of millions of people will starve to death in spite of any crash programs embarked upon now." Ehrlich was flamboyantly wrong. The green revolution intervened. Since

1968, famine has in fact declined sharply around the globe. But it is perfectly fair to reply, as Ehrlich now does, that these issues are too important to be reduced to small matters of timing. The starving masses will be no less miserable if the mass famine comes a generation later than Ehrlich expected. Nor will the ruined planet be any less ruined by their desperate struggle to survive.

The Hard Green directs his scorn next at the Club of Rome and its 1972 study, *The Limits to Growth*.[2] Humanity was scheduled—almost as precisely as a Swiss train—to run out of gold by 1981; mercury by 1985; tin by 1987; zinc by 1990; petroleum by 1992; and copper, lead, and natural gas by 1993. The Club was wrong, too: The average price of all metals and minerals in fact fell by more than 40 percent between 1970 and 1988. And oil is cheaper than ever before.

Still, as the Softs are forever reminding us, the final accounts aren't in yet. In the past century we have burned up fossil fuels that took millions of years to create. We shall burn as much again, and more, in the next. The rain forest is being leveled, the seas exhausted. Spendthrifts who inherit a great fortune are still spendthrifts, even if they discover halfway through a life of profligacy that they inherited even more than their doleful accountant had previously supposed.

MALTHUS ON A CHIP

In a modern economy, Malthusian accounting requires more than ordinary mathematics. Malthus himself made a fairly elementary mistake, after all. To grow more food, you can add more oil instead of more land. Oil, packed with (fossilized) solar energy, can be used to produce fertilizer, which makes crop yields soar. The modern, global, industrial economy allows tens of thousands of other trade-offs like that one. With so many different stocks in the portfolio, it takes more than a doleful accountant to keep track of the spendthrift. It takes a doleful computer.

Environmentalism's descent into silicon can be traced to a man who developed the first large digital computers in the 1950s to track Soviet bombers. MIT electrical engineer Jay Forrester had been investigating the use of magnetic materials to store digital information when he joined the team that was designing an early computer, called "Whirlwind," for the Navy. The initial objective was to build a machine that could analyze airplane stability and control. Forrester redirected the project to design a general purpose digital computer. As the machine evolved, the Air Force proposed a new mission for it. Forrester's computer would receive data from dozens of radar operators, store and instantly process it, and then vector defensive fighters to attack inbound Soviet bombers.

With the help of a brilliant graduate student, Ken Olsen, Forrester assembled the first truly functional digital machine. Tests conducted in May 1952 were stunningly successful. Forty-eight aircraft "attacking" Boston were accurately tracked and intercepted. IBM was brought in to transform the prototype into the nation's SAGE air defense system.

In 1955, Forrester moved to MIT's newly opened Sloan School of Management, in search of a new challenge. As he searched, IBM built increasingly powerful computers. Herbert Grosch, a one-time IBM employee, demonstrated that a computer's power would increase with the square of its size. "Grosch's Law" meant building fewer, bigger machines. The entire world, IBM estimated, would end up being served by about fifty-five big mainframes.

By the 1960s, however, both the uses and the markets for mainframes had already expanded. MIT itself owned one. At Sloan, Forrester set about using it to model industries and then cities. From a computer's perspective, the problems were not all that different from tracking bombers. Both come down to simultaneously solving large numbers of nonlinear differential equations efficiently and accurately. Forrester had mastered the altogether new skill of moving otherwise unsolvable equations into the new digital machines. In 1969, a year after Ehrlich's *Population Bomb* was published, Forrester published *Urban Dynamics*, a book about the functioning of large cities.

A year later a group of policy makers, academics, and managers calling themselves "The Club of Rome" convened in Bern, Switzerland, to discuss hunger, pollution, and other "world problems." Carroll Wilson of MIT, a member of the Club's executive committee, brought Forrester along. On his way home from the meeting, Forrester sketched out a rough model of planetary resources, which he called "World1," on the back of an envelope. He promptly transformed it into a large computer program. Release 2.0 (World2) would follow, then World3, then *World Dynamics*.

Forrester's models were Thomas Malthus brought back to life, torrents of gloomy electrons in the solid-state brain of a machine. The mainframe computer was critical; the model could never have been solved without it. Forrester's model consisted of some forty-five inter-connected subsystems. Typical subsystem blocks were NRUR (natural-resource-usage rate), DR (death rate), POL (pollution), CID (capital investment discard), BR (birthrate), and so forth. Agricultural investment increased agricultural output, which increased birthrate but also pollu-tion; pollution decreased agricultural output; and so on. Out of the model emerged predictions of things like total world population, total pollution, and quality of life. Quality of life, the model indicated, had peaked in 1940. With the computer script already written and filmed, the book was easy. Published in 1972, under the auspices of the Club and Forrester's acolytes Dennis and Donella Meadows, *The Limits to Growth* would be translated into twenty languages. It sold nine million copies.

The Limits transformed the fuzzy concept of scarcity into the holy digital writ of the Soft Green movement. The book purported to calcu-late what would ultimately "limit population and physical growth on this finite planet, and how the world's adjustment to its limits might be smooth and controlled rather than unexpected and violent." The predicament of mankind was not good, *The Limits* concluded. The world was rapidly reaching the end of its ecological tether. The model the book described predicted a series of imminent disasters: vast upswings in population punctuated with massive die-offs and steady decline in the quality of life. A "rapid, radical redressing of the present unbalanced and dangerously deteriorating world situation is the primary task facing humanity."

The book's timing could hardly have been better. The Arabs embar-goed their oil a year later. The endless gas lines that ensued seemed to vindicate Forrester completely. Forrester's methods endured and grew. And thus it was that Malthus and the digital machine converged.

THE RISE OF THE MODEL

The models have grown ever since, as fast as the computers on which they run. Year by year, they grow more detailed, complex, and graphic. Indeed, the computer modelers might well ask at this point a completely self-referential question: How much life can they pack onto half a square inch of sand? Experts have been pondering that question for over thirty years. The sand is silicon; the life, software; the habitat, a microproces-sor. The size of a processor hasn't changed much since the first one was built in 1971. But the population within has grown geometrically, year by year, from tens of transistors in the 1960s to tens of millions today. Moore's Law, set out in 1978, predicts a doubling of the number of transistors on a chip every eighteen months. Gordon Moore, meet Thomas Malthus.

With the help of a sufficiently powerful computer, a numerical model can be built for anything: a bridge, a space shuttle, a city, a war, or a planet. Every modern designer of a car, radio circuit, or the control surfaces of a wing analyzes the system by building a mathematical model of the component pieces and then numerically analyzing how the system as a whole will behave. This is "system dynamics," a well-established, quantitative discipline for tracking flows of electric current, fluid, heat, and mechanical energy through a system, to determine how it will move, warm up, cool down, oscillate, resonate, amplify, stabilize, or die out. The basic rules are not very different from rules of accounting: Track energy and material as they move through the system, conforming with conservation laws that are familiar and well understood. The complexity lies in the number of connections, not in the basic principles.

Big computers are quite capable of keeping track of lots of connections. As Forrester's electronic progeny grew in the 1970s, it was reasonable to suppose that the machines might be adapted to track the stability or collapse of anything that is bought, sold, or driven; that freezes, boils, or melts; that vibrates, decays, or collapses. More complex systems just meant more equations, which simply required bigger computers. IBM would take care of that end of things.

So if we can model a car on a computer, why not the resources of a planet, too? The basic rules seem straightforward enough. Humanity resides on a finite surface, the planet itself. There is only so much dry land; there can be only so much oil beneath it. Supply and demand are matters of basic bookkeeping. Production and consumption must balance. Boeing designs new jets using similar models: Small quantities of basic science run through large engineering computers. If you just don't trust big computer models at all, never set foot in a jumbo jet.

Soft Greens, distrustful of high technology in all other arenas, have not hesitated to trust it here. Their computer models have become hugely influential in modern environmental discourse. Backed by a model, almost any grain of research, no matter how tiny the scientific backyard in which it takes root, can grow to full-planet size, with commensurately grand implications for public policy. It is the model that generates most of our environmental headlines today: species extinction, global warming, and almost every new cancer scare. It is the model that proves that however plentiful things may look just now, we will in due course run out: of rice, wheat, or food in general; of copper, zinc, and other raw materials; of new genetic stock for medicines and agriculture; of gas and oil; of land to live on; of dumps to contain our copious wastes. Feed a minimum of data into a maximum of computer code and then let the machine whirl off in time and space. The models let you think locally, pronounce globally.

For academics, pundits, and policy makers, the model has become as important as the "Earthrise" poster is for the rest of us. It lets us see it all, in cosmic perspective: the blue and white sphere of the Earth rising over the stark white horizon of the moon against the blackness of cosmic space. The models purport to reveal the ecological planet suspended ever so delicately, in both time and space.

Models and Markets

The Limits modelers were smart people, equipped with the finest computers of their day. These were the people and the machines we were counting on for our national defense. Forrester had a brilliant track record. His protégé Ken Olsen went on to found the Digital Equipment Corporation.

So it is surprising to find just how dismally *The Limits* models performed. Global economic collapse was supposed to have arrived a decade ago; what arrived instead was a global economic boom. They predicted the exhaustion of everything; we see instead glut upon glut. The people entrusted to guide our fighters to intercept Soviet bombers proved unable to guide your Buick to a fossilized tree. They predicted we would have run out of oil by now and our wheels would be up on bricks. You have in fact traded up to a monstrous Chevy Suburban, and you drive it more than ever. How did such clever people, such powerful computers, manage to get things so very wrong?

In retrospect, it is perhaps tempting to dismiss anything called a "world model" as the work of a crank. The label itself brings to mind one of those eccentrics who occasionally buys space in a newspaper, where he claims to set out in a single, very elaborate equation, the inner workings of everything, the secret of universal harmony, and the key to world peace. But Forrester's methods were not the methods of cranks. They were the methods of people who successfully design jets and coordinate our national defense.

It is equally inadequate to argue that the world is just too big to model. Modeling the flight of the world is, in some respects, easier—not harder—than modeling the flight of an airplane. Big things, and overall trends, are often easier to predict than finer details. It is easier to predict where the planet itself will be six months from now (that is, on the other side of the sun) than it is to predict just where a jet will be thirty seconds after an engine falls off. Next July will be warmer than last January; the weather next week is a tougher call. The fundamentals that Forrester set out to track—materials, energy, populations, and so forth—are all ultimately governed by the basic physical constraints of the planet, which should make the modeling easier. The Earth, its atmosphere, its oil reserves, its land mass, are all finite. Surely it is possible to crunch the numbers and arrive at some overall limits on how big a biological cake can be baked with these ingredients.

It is indeed possible; it just isn't useful. On the first cut, the cake turns out to be ridiculously huge. By global standards, life occupies only a very thin film on the surface of our planet. Moreover, atoms are indestructible: We have that on the authority of Democritus of Abdera, who so named them ("atomos," "indivisible") in 430 B.C. So one cannot "exhaust" the Earth's supply of copper; the worst one can do is disperse it. And while energy can be degraded, humanity uses only a minuscule fraction of the vast amounts of nuclear energy continuously released in the core of our own planet or directed toward us after its release in the core of the sun. Just looking at the standard things normally tracked by "system dynamic" models, there's enough matter and energy out there to sustain and replicate humanity billions of times over.

It is at this point in the standard Hard Green tract that one normally sets out some illustrative figures. Standing shoulder to shoulder, all the people of the Earth would scarcely fill Delaware. All our copper, tin, zinc, and deuterium could be culled from one-billionth of the world's oceans. All our energy requirements for two thousand years could be supplied by one-quadrillionth of the thermal energy produced by radioactive decay in the core of the Earth. Suitably stacked and compressed, all our trash would fit in a pyramid just 2 miles high in southern Connecticut. All our drinking water could be supplied by the rainfall of Oregon. Where did I get all these numbers? I made them up. I could have dug up real ones from the extant, Hard Green literature, and they would have read much like the ones I invented. But why bother? People aren't going to live shoulder to shoulder in Delaware, or build pyramids in southern Connecticut. Exxon knows where to find gasoline, not deuterium. Anaconda knows how to extract copper cheaply from a mine in Utah, but not from the sea. Forget about it.

So that is what Forrester did. He assumed the basic constraints defined by existing technology, markets, and patterns of supply and demand, as they existed in 1970. Not the speculative far future, not pie-in-the-sky, just the basic, sensible, practical present. He based his model on known technology, known mines, known reserves, known quantities of arable land, and known rates of growth in all of the same. From the get-go, Forrester had to shrink his model down to size: not down to geophysical size, but down to *human* size. To escape the big-Earth problem, Forrester shrank it, right down to the size of an academic's mind at MIT in 1970.

This is what practical-minded engineers always do. If you are designing a jet you assume more or less current technology, current engines, current materials, with cautiously optimistic allowance for improvement and innovation here and there. Oh yes, you make some allowance for technological innovation—whatever you can reasonably imagine coming down the pike of your discipline some time soon. You allow for modest variations in geometry. To model the flight characteristics of a caterpillar, you model the worm you see, not the butterfly concealed within.

But like the caterpillar, people change, even more than a caterpillar, in the one dimension that matters the most. To simplify the rest of the story only a bit, one might say that Forrester was not smart enough to model the future ingenuity of intellects like Ken Olsen, a protégé of Forrester himself. Olsen had completed his studies at MIT in 1952 and had gone on to found the Digital Equipment Corporation in 1957. By 1972, when *The Limits* was published, Digital was a $188 million a year company, with all but limitless prospects out ahead of it. Worse still for *Limits* modelers like Forrester, a new generation of wildly creative young Olsens, of equally unlimited talent and ingenuity, had enrolled at MIT.*

Forrester counted mouths, but behind every human mouth there cogitates a brain. That was how Julian Simon would later see it. Simon, an anti-Malthusian academic at the University of Maryland, would respond to the *Limits* crowd with a series of books and articles declaring all neo-Malthusian predictions to be bunk. He dispensed with models, examined historical trends, and concluded that growing population makes people richer, because each new brain more than offsets each new mouth. As-

---

*I knew some of them; I was an MIT sophomore myself.

sume that, no more, and you turn all Malthus, all *The Limits*, upside down. Simon understood that mouth-brain trade-offs are a lot more subtle than that. But however simpleminded my own summary of his voluminous work may be, it is less simpleminded than the one contained in Forrester's labyrinthine computer models. Forrester made no real allowance for extra brains at all. He never imagined that pounds of sand might soon displace tons of copper. Corning Glass did, and gave us fiber optics.

Innovators have conspired to thwart Malthusians since the beginning. Food caught the interest of another cleric born in Malthus's lifetime, who spent his time cultivating peas in the garden of his monastery. Gregor Mendel thus learned the science of genetics. His intellectual descendants, the bioengineers of the green revolution, devised strains of high-yielding cereals that made nonsense of all Ehrlich's prophecies of famine. One might have supposed that the mentor of a man like Ken Olsen would make some allowance for genius in his models, but Forrester couldn't. An engineer can model a car, but he cannot model another engineer, least of all a smarter one. The collective genius of humankind cannot be contained in a model.

The closest we come to getting a grip on human ingenuity is in the marketplace, which is not a model at all; it is a process. The market is the place where innovation intersects with labor, resources, pollution, and quality of life: all the other things Forrester was trying to track. Most of the intelligence that traffics in the market falls well short of genius, but in the aggregate, it is ordinary intelligence that matters the most. Free markets elicit and distribute information and ingenuity in ways no other process can match. Forrester made no serious allowance for market forces at all.

He couldn't. If markets could be reliably modeled, as the Soviets thought they could, we wouldn't need markets at all. We wouldn't want them: Competition has many disadvantages, including many irreducible inefficiencies. We put up with markets only because, as bad as they are, they are the least bad economic option around. They discover their own future, far more efficiently than any model can predict it. Indeed, no model has ever been developed to predict tomorrow's stock prices reliably, let alone next year's. Lots of people peddle such models anyway, models as reliable as a dowser's rod or the crystal paperweight of a soothsayer. The most fundamental lesson of economics is that markets as a whole "know" things the rest of us don't. The "efficient market theory" establishes, quite conclusively, and with a wealth of empirical evidence to back it up, that the market always "knows" more than any player in it. Modeling markets is not difficult. It is impossible.

But recall that Forrester needed the market—at the outset—to deal with the big-Earth problem, to shrink the Earth down to size so as to discover some interesting limits in his models, some numbers smaller than "billions upon billions." He relied on the market to tell him what was possible in 1970, then relied on his computer to tell him what would be impossible in 1997. He assumed the market to construct his model, then assumed *no* market to run it, and ended up proving that the market would fail. He presumed, in other words, to predict the future of what he assumed did not exist.

THE LIMITS OF MODELS

Forrester taught us nothing useful about the limits to growth, but he did teach us something useful about the limits of models.* To begin with, dead reckoning is impossibly hard. Forrester's Whirlwind directed the fighters to the bombers but only because it was richly fed with constantly updated information from the real world. Absent such feedback, mistakes pile up. In the real world you can generally see the drift as soon as it begins and constantly correct it. But a feedback-free model has to start right and stay right as it whirls off alone in space and time.

Forrester also taught us that the biggest computer can't make up for critical deficiencies in the model itself. Forrester left out markets and innovation, roughly like a biologist trying to model the history of life on Earth without allowing for mutation or survival of the fittest. Finally, and most fundamentally, Forrester proved once and for all that some essentials—like markets and innovation—can't be modeled at all. They are way too subtle and complex for that.

This is where even the most sensible observers often go wrong: They simply forget to take account of everything we don't yet know. In her 1992 book *Costing the Earth*,[3] Frances Cairncross, the *Economist*'s environmental editor, cogently argues that some forms of macro-economic "growth" are simply products of dishonest accounting. Rapid "growth" in the Philip-

---

*Though some people never learn. The authors of *The Limits* issued a sequel in 1992, titled *Beyond the Limits* by Donella H. Meadows, Dennis L. Meadows, and Jørgen Randers (White River Junction, Vt.: Chelsea Green Publishing, 1992).

pines, achieved by the one-time harvesting of over 12 million acres of hard-wood forests, resulted in denuded hillsides, silted rivers, and declining food production and fish catches. Macro-economic bookkeepers often neglect to keep books for "intangible" assets like environmental quality.

True enough, but those same bookkeepers invariably neglect to keep books for the intangible of intellectual property as well. Even if they re-member patents, which are almost impossible to value in any event, they cannot add the sum total of what a populace has learned. But that is where an ever-growing fraction of national wealth resides. Which means that macro-economic books are of no use for all the things that really matter in the long term, environmental and intellectual alike. The most we can say is that for the last two centuries at least, intellectual gains have surpassed environmental losses on the missing ledger of intangibles.

Tellingly, the people who use computers to model everything else can-not begin to model the trajectory of the computer itself. Herbert Grosch had a model for the mainframe, and it was good for a while, but then it was knocked on its head by Moore and the silicon microprocessor. Today, IBM maintains that it is inefficient, even self-destructive, to keep packing more gates into a microprocessor. Eighty percent of a microprocessor's instructions are executed by 20 percent of the gates. A "reduced instruc-tion set" chip is faster and more efficient than the alternative, the "com-plex instruction set" chip. RISC beats CISC, it is said. Control the population (of gates) and life (on the chip) will get better. Similar debates rage higher up in the cybernetic food chain, between purveyors of fully equipped desktop machines ("fat clients") and stripped down "network computers" ("thin clients"). And higher still, between fully loaded oper-ating systems (Windows) and stripped down alternatives (Java).

Like the elephant on the Serengeti, the microprocessor shapes an en-tire habitat for others, the dung beetles and the Baobab trees alike. Life in the microcosm exists on a silicon plain of our own design and manu-facture. We can dam and dike, fence and gate, build or raze, etch, burn, irradiate with X-rays, and lithograph with impunity. We can create new species and wipe out old ones. The EPA will not interfere. We are both the Creator and the Destroyer of this world, its only Darwinian avenger. But for all that, we have no real idea where it's all headed. The best minds, the wisest cyber-pundits, the wiliest NASDAQ stock pickers, the highest-flying venture capitalists, do not even begin to agree.

Modeling the potential resources of a planet is a lot harder than mod-eling the potential resources of a sliver of silicon. Much harder, for the simple reason that the resources of the planet are defined and extended by the power of silicon itself and the billions upon billions of infinitely smarter human neurons that function behind it. Smart as they were, the *Limits* modelers couldn't come close to modeling all the rest of human intelligence. No one ever will.

The Malthusian era can be said to have ended officially on October 11, 1990. Anti-Malthusian Julian Simon had kept on writing and writing. First, the neo-Malthusians dismissed him as a lunatic. Finally, to shut Simon up, Paul Ehrlich accepted Simon's 1980 offer to stake $10,000 on long-term futures "in any standard mineral or other extractive product you name." Ehrlich got to choose both the commodities and the 1990 day of reckoning.* He ended up losing $576.07 to Simon in the Malthusian trading pits.

LIMITS, REGARDLESS

Free markets beat modelers every time. The inflation-adjusted price of raw materials has, with only a few, comparatively modest interruptions, been falling steadily for two centuries. For any good that is shadowed by paper on exchanges in New York or Chicago—all capital, all labor, all ordinary services like transportation or health care, everything traded or sold in normal markets—predictions of pathological disequilibrium invariably fail. People simply don't run out of things they can package as "property" and trade freely in unregulated markets. With markets in command, scarcity is always giving way to abundance. Forrester and the *Limits* modelers didn't get the story even half right. They assumed no market, from which they concluded the market would fail and we would all get poorer. But there is a market. It does not fail. We keep getting richer.

And yet, however wrong Forrester may have been, who can really doubt Malthus? Anything that starts growing geometrically will have to stop growing at that rate, one way or another, sooner or later. The growth can stop soft or stop hard, stop by catastrophe or by collapse; it can stop with a bang or a whimper, but it is going to stop. One can argue

---

*Ehrlich and two of his colleagues invested $200 in each of five metals: chromium, copper, nickel, tin, and tungsten. They chose September 29, 1990, exactly ten years later, as the end date. If the inflation-adjusted prices of the chosen metals rose during that time, Simon would pay Ehrlich the difference. If the prices fell, Ehrlich would pay Simon.

timing, but no more. The laws of mathematics, at least, are beyond dispute. Geometric growth ramps up toward infinity. But infinities don't happen in the real world.

So what then *will* impose the ultimate limit to growth? Will we choke on our own wastes? No. Will our high technology, our nukes, and genetically engineered strawberries crash down upon us? No. Can environmental salvation lie instead in making ever more efficient use of existing resources? No.

As I argue in Chapter 9, the real limits to growth lie elsewhere. It is reasonable to hope that they will save humanity from Malthus. And that they will save the rest of nature from humanity, too.

2

Externality:
Pollution on a Chip

———— ◆ ————

For five centuries of common-law jurisprudence, they were just called "nuisances." The term covered injurious or obnoxious things that didn't quite crystallize into assault or trespass: smoke; vibration; noise; sewage; and the keeping of dangerous wild animals, explosives, gaming houses, or harlots where they didn't belong. The "nuisance grounds" was the town dump.

In 1995, Julian Simon rashly suggested that Paul Ehrlich accept a second bet along lines similar to the first. Ehrlich promptly offered to stake $1,000 per trend on fifteen "continental and biological indicators."* In the global atmosphere, temperature, carbon dioxide, nitrous oxide, and ozone would all rise, as would sulfur dioxide emissions in Asia. On the surface, fertile cropland, agricultural soil, firewood, virgin tropical forest, oceanic fisheries, rice and wheat production, and plant and animal species

---

*All comparisons were to be between conditions in 1992–1994 (or just 1994) and those in 2002–2004 (or just 2004): (1) a warmer planet; (2)–(4): more atmospheric carbon dioxide, nitrous oxide, and ozone globally; (5) more sulfur dioxide emissions in Asia; (6)–(7): less fertile cropland, and less agricultural soil, per capita; (8) less wheat and rice grown per capita; (9) less firewood per capita in developing countries; (10) less virgin tropical moist forests; (11) lower oceanic fisheries harvest per capita; (12) fewer plant and animal species surviving; (13) more people dying of AIDS; (14) lower sperm counts in human males and more reproductive disorders; and (15) a wider gap in wealth between the richest and poorest 10 percent of humanity.

would all decline. AIDS deaths would increase and sperm counts in human males would decline. And the "gap in wealth between the richest 10 percent of humanity and the poorest 10 percent" would grow.

The first bet, the one Ehrlich had lost to Simon, had been all about things that are traded in markets. Ehrlich carefully framed the second around things that aren't. Simon wouldn't touch any of those. He would put his money only on "direct" measures of human welfare, such as life expectancy, leisure time, and purchasing power.*

NUISANCE

Markets don't take proper account of certain costs, smoke and sewage among them. Economists give the idea a formal label, then restate it as a tautology: Markets don't take proper care of "externalities." The man who produces the pig does not in fact make good use of every last part. Not the squeal, not the smell, not the manure. He dumps them in the air and river instead. We are assailed, as a result, by smoke, sewage, and trace chemicals, by noise and vibration, by eyesores and rampant ugliness, by radiation and microwaves, all unloaded into spaces that belong to none or to all. The Soft Green can extend the list indefinitely, to include even the dump itself. In 1987, when the *Mobro 4000*, a barge loaded with Long Island trash, had some trouble dumping its load, got on the national news, and then wandered aimlessly for thousands of miles, America followed.

· To put the same case in terms of goods rather than bads, markets rarely attach proper value to public goods: wolf and forest, eagle and ozone layer, whale and ocean. Soft Greens often cite a 1968 parable published by ecologist Garrett Hardin titled "The Tragedy of the Commons."[1] The traditional village commons was a public pasture, open to everyone's cattle. Every villager had a personal incentive to use it fully, so the herds kept growing until once good land was entirely destroyed.

And while all the details remain to be debated, the Soft Greens are simply right. They have here an unanswerable case for government intervention of some kind. The intellectual heights are so commanding that Soft Greens scarcely bother to fortify them any more. The triumph is so complete that the Softs do not even read the enemy's articles of surrender. "Many popular textbooks on economic theory fail even to address subjects as basic to our economic choices as pollution," Vice President Al Gore con-

---

*The betting is over now, at least between those two sportsmen. Simon died in 1998.

fidently declares in his 1992 manifesto, *Earth in the Balance*.[2] As MIT economist Paul Krugman points out, it hardly matters that in fact, *every* popular textbook prominently discusses pollution as the *paradigmatic* illustration of how free markets sometimes fail to get things right. Al Gore journeys to Kyoto, to express his solidarity with the global atmosphere, the most external of all goods that every economic textbook he reads fails to notice. All but unnoticed, themselves, Hard Greens are reduced to quibbling about whether global warming is really quite so bad as the Softs maintain.

There is no more talk of Malthusian starvation here. If humanity were destined to run out of food quite soon, we wouldn't have to worry much about its wastes. Starving people do not fill up new dumps, they set up shanty towns around old ones, to rake the muck, sift the trash, and feed on the refuse. Throw-away is the mind-set of abundance, not scarcity. By losing so convincingly on scarcity, however, Soft Greens start all the more strongly on pollution. With pollution, the Softs advance boldly, under the flag of principle. It is the Hards, now, who nitpick the terms of their surrender.

The Hards will have to get used to it. Progress has been made with some of the worst externalities: big smokestacks, car emissions, industrial effluents, sewage. But there is still much to be done. Few economists of any stripe believe we should actually be doing less to protect the environment. Proper protection of the environment requires more of something or other: more regulator, tax collector, or licensing authority, perhaps? The Softs find themselves in the delightful, table-turning position of demanding more accountability closer to home, more leave-us-alone principle, more secure walls around your private factory (and its smokestack), your car (and tailpipe), your barbecue (and its smoke), your trash bag (and its dump). The Hards wonder where their own principles have gone and find themselves pleading for a little more tolerance and a little more willingness to share the rough with the smooth.

So why is it taking so long to get things properly in hand? Krugman blames public ignorance, interest group pressure, and the political power of the reactionary Right. We are just too politically stupid to correct the environmental externalities that assail us.

But it is not quite as simple as that.

EXTERNALITIES UNLIMITED

As we saw in Chapter 1, the first problem for the Malthusian analyst is that markets have a future, and a good one, if experience is any guide.

The first problem for the Nuisance analyst is that markets have a past, and a bad one.

Pollute though they will, new cars, water heaters, air conditioners, and power plants don't, for the most part, add; they replace. And because technology is steadily improving, the replacement process is the best thing we have going to improve the environment. Tolerating a little new smoke is often the most practical way to rid the landscape of an older chimney that is much dirtier.

There are similar, equally cogent arguments for tolerating some very big polluters. The large centralized power plant and factory are easy to regulate. But per unit of output, they are often much cleaner than the decentralized, low-tech, historical alternatives they displace. It is far more efficient, for example, to burn oil in the huge, meticulously maintained boiler of a central power plant than to burn it in a two-stroke engine of a lawnmower, even after we allow for all additional losses in transmitting electricity from the power plant to the end user. Soft Greens understand such trade-offs well enough when the debate is whether to favor cars or trains. Here again, the real enemy is not pollution in the abstract, but how to make incremental improvements where best we can.

Across the economic landscape, pollution trade-offs are a lot more subtle and complicated than Softs generally allow. Disposable paper diapers can claim to be more environmentally friendly than reusable cloth because they don't have to be washed in hot water. Styrofoam cups insist they are greener than ceramic mugs, for similar reasons: less energy used in both the making and the using. Plastic packaging maintains it is kinder to the land than paper because it uses less material and reduces the spoilage of food more effectively. New cars excuse their own pollution on the theory that they pollute much less than the old clunkers they displace. Even garbage dumps can mount a spirited defense of their greenish bona fides. They accuse recyclers of consuming more resources from the trucks and sorting plants and such than they save. Recycling newsprint, they charge, creates more water pollution than making new paper. We need not buy all of these accounts, but we must at least weigh them very carefully. They may be true.

Many of them fly in the face of what has become conventional environmental wisdom. Having been assailed the longest and hardest, the nukes are by now the undisputed masters of self-apology. Nucleo-phobes are stunned to learn that nuclear plants are less radioactive than coal, which contains trace radio-nucleotides that go straight up the stack. Less radioactive than natural gas, too, the fossil fuel Soft Greens dislike the least: Gas contains traces of radon, which entails the release of more radiation into the environment when gas is burned in a power plant and which creates comparatively huge additional exposures to radiation when gas is burned in the home. And while natural gas releases less carbon dioxide than coal when burnt, the methane in the gas itself is a far more powerful greenhouse gas and some of it leaks. Uranium watts produce no greenhouse gas. With breeder reactors propelling its nuclear-electric economy, France lectures America on the unacceptability of burning coal.

The details are debatable; not debatable is that such trade-offs stand at the center of all serious environmental debate. There are difficult, second-best choices to be made, between old pollution and new, between centralized power and decentralized, between public transportation and private. As a practical and political matter, the market's history makes market externalities very much more difficult to address than economists like Krugman acknowledge. Rules that might have been perfectly efficient if announced long before the market arrived may not be efficient at all if announced after the fact. But the rules cannot be announced in advance because the market is always in the process of arriving. The future of the market is unknowable. So too, therefore, is the future of the market's externalities, its pollution.

And before there was market there was nature itself, and it too has a history, a Darwinian one. The makers of synthetic pesticides have the temerity to argue that pesticides actually reduce health hazards: They substitute for the all-natural ones otherwise produced by plants that have been bred to resist pests on their own. The styrofoam cups and plastic forks boldly claim to defend humanity against externalities of nature's own creation. Bacteria levels on porcelain cups and flatware in restaurants are hundreds of times higher than on disposable ones. Do the Soft Greens dare to stand up and defend E. coli and Salmonella and the essential roles they play in the overall ecological balance? Sitting as jury, the ordinary American family may be unwilling to extend its green sympathies quite that far. Intestinal bacteria aren't produced directly by the market to begin with, so they are not technically speaking "externalities" as defined in the economist's criminal code. But for the person choosing between a plastic fork and metal one, they are part of much the same anti-social racket.

The problem here is that however eager we may be to reduce our own externalities, the rest of nature is not the least bit eager to return the favor.

Nature is not inherently benign; still less is it predisposed to favor our species over others that aspire to consume, colonize, or digest us. Nature doesn't contrive in favor of any particular species; to suggest otherwise is to deny Darwin. Chiggers, fleas, lice, and tapeworms do not bid for blood-sucking permits on the Chicago exchange. So far as nature is concerned, humans are habitats themselves, on or within which other species aspire to be fruitful and multiply. It was indeed notable to find AIDS in the list of "continental and global indicators" chosen by Ehrlich in the follow-up bet he proposed to Simon. An "environmental" problem? Well, yes, in fact it is. But not one likely to be solved by directives from the EPA.

This is why humans create so many environmental "externalities," both consciously and deliberately. Like all other creatures of evolution, humans engage in ceaseless, deliberate, and sometimes genocidal warfare against other living creatures that assail them; the only real difference is that we fight with our brains, too, not just our claws and our antibodies. We vaccinate to boost those defenses, sometimes (as with smallpox) to species-destroying levels. We use plaster in our walls because it stays so dry that nothing can live in it. We scorch cockroaches any way we can. Fighting back against those parts of nature that assail us is an externality-creating enterprise, quite as much as the assailing itself. There are, of course, better and worse ways to mount our defenses against nature, but there will be defenses, and they will, by definition, disrupt the "natural" environment. Here again, there are no truly green choices at hand; some are just less anti-green than others.

Nature, too, thus makes the externality problem far trickier than simple-minded economic theory suggests. The omniscient, cost-internalizing economist-in-the-sky can't address most externalities at all, he has to leave them to your immunologist and your exterminator. So far as externalities are concerned, the market is really the last and least of humanity's problems. Nature itself decrees that there is far more to the "externalities" problem than first meets the economist's eye. Things casually denounced as environmental vandalism can begin to look, upon further reflection, like civic duties. The Bureau of Soft Green denounces the plastic fork. The Department of Health requires it.

At the end of every externality debate stands a single, incontrovertible truth. All forms of life take in higher grade energy and emit lower, take in material in one form and emit material in another. Energy flows up the food chain, increasingly concentrated, from sun to grass to rabbit to fox, from algae to whale. Material flows with it. We drink and eat, sweat, smell, and excrete; we ovulate, ejaculate, and give birth. Any one body can be recycled perfectly, dust to dust, ashes to ashes. But life as a whole does not recycle. It cannot. If it is semantically ridiculous to say that "trees pollute," that is simply because "pollution," like "externality," has been defined to refer to humanity alone. But semantics are beside the point. A river of matter and energy flows through every living thing. It never emerges in the same form as it enters.

The most fundamental laws of physics require this to be so. Assembling a chicken from an egg entails an increase in biochemical order. Normally, things head the other way, from higher order to chaos. The Second Law of Thermodynamics requires that, overall, they must. An increase in order—a decrease in entropy—is never possible in a completely closed system. Life on Earth is possible only because we happen to sit in a thermodynamically sweet spot between hot sun and cold outer space. Heraclitus was right. No man can step twice into the same river.

None of this proves you should prefer a plastic fork to a well-washed metal one. The laws of thermodynamics do not compel us to gorge on fat or to race power boats through the Everglades. But at the end of all such debates, one stubborn fact will remain: Life cannot be contained. The choice is not between life with externalities and life without. Life is an externalizing state of being.

POLLUTION ON A CHIP

So the externality debate ultimately pivots on questions considerably more subtle than is often acknowledged. The debate is always which externalities to pursue, in what order, and just how far to pursue them. There is a limitless list of candidates and no natural limit at all to how far the pursuit might be extended.

Many of the candidates are obvious and the need for improvement is so clear that there is no need for extended debate. The legacy nuisances of yesterday's practices are often the biggest and easiest to see, through today's eyes. Trash. Sewage. Pea-soup fog. Horse manure piled high in the city streets. Nobody needed a model to explain such things or what was wrong with them. Pastoral Romantic might debate Industrial Gradgrind as to whether they were a necessary and acceptable price of progress. But neither doubted they were part of the price, and a serious one.

The wealthier we grew, the less acceptable these nuisances became. In the 1960s and early 1970s, states and the federal government enacted laws to curtail pollution of air and water and to promote sanitary disposal of solid waste. As I discuss in Chapter 8, the job could have been done much better, faster, and cheaper by other means, but be that as it may, at least some of the laws we got are doing the job, too. At the level of the large and visible, things have improved steadily ever since. Our society has grown adept at limiting the visible forms of pollution and their immediate, short-term impacts on environmental quality. The problems are far from solved, but they clearly are on a steady path of improvement.

But the more we have abated forms of pollution that are both visible and unpleasant, the more preoccupied we have become with other forms that are neither. Spatially, modern environmentalism focuses more and more on the extremely diffuse, the vanishingly small trace of dioxin or pseudo-estrogen or lead. Temporally, the focus is even more on the very long term, the cumulative impact of acid rain on several decades of forest growth, the effect of atmospheric carbon dioxide on global temperature over the next century. At least seven of the fifteen "indicators" Ehrlich proposed to Simon for a second wager involved things that no ordinary person could ever sense or track. Such "pollutants" require the most delicate forms of measurement. Their impacts on human or environmental health—if they have any such impacts at all—can be discerned only through a fog of computer models and statistics.

In the spatial dimension, the models revolve around epidemiology or its ecological equivalent. With air quality improving steadily and the main aesthetic goals of regulation in reach, the EPA now labors to prove that industrial air pollution causes sudden infant death syndrome (SIDS). "High soot levels increase risk of babies dying," the headlines declare. It is suggested that certain pesticides are pseudo-estrogens that may disrupt human hormones and depress sperm counts in men or increase breast cancer in women.* Medical records are combed, sperm banks are sampled, data are gathered from far and wide, and then the computer takes over. The most elaborate statistical analysis is required, because there are usually dozens, sometimes hundreds, of possible confounding variables.

---

*The sperm count crisis was the main thesis of the 1996 book *Our Stolen Future: Are We Threatening Our Fertility, Intelligence, and Survival? A Scientific Detective Story*, by Theo Colburn, Dianne Dumanoski, and John Peterson Myers (New York: Penguin Books, 1996). Ehrlich, always up to date, chose sperm counts as one of his fifteen indicators in the 1995 follow-up bet he proposed to Simon. One might, however, have supposed that lower human fertility would be welcomed by such strong and long-standing advocates of population control.

The models crunch reams and reams of numbers in search of an association between the one suspect and its supposed consequence.

In the temporal domain, the ratio of computer to data is higher still. Carbon dioxide is known to act as a "greenhouse gas," at least in laboratory conditions. Levels of atmospheric carbon dioxide have been rising for a century, so how hot will the Earth be a century from now? Even more elaborate models crunch even more numbers and emerge with a prediction.

Much of the time, the last step in the modeling is to explain why the harm, in the end, will be to humans themselves, not just to the environment around them. Environmental activists finger PCBs as the paradigmatic industrial toxin to target in the Toxic Substance Control Act of 1976 because they are said to cause cancer and birth defects. Federal legislators, *The New York Times* reports, conclude that banning PCBs is "a moral responsibility to the people." In 1978, three federal agencies declare that occupational chemicals cause at least 20 percent of human cancer. Others conclude that the pesticides we use to kill insects will end up killing us. The carbon dioxide that warms the globe will lead to a great plague of mosquitoes and epidemics of dengue fever and yellow fever in North America, predicts the "Hotter Earth, Madder Mosquitoes" model.

And so the environment descends for the second time into the recesses of the computer chip. In their basic conception and operation, the models themselves are all much like Forrester's. They have the same ambition, the same vast sweep in space and time; they run on the same computers. But the context now is quite different. The models are now about bad things outside the market, rather than good things within it.

At first blush, this would seem to make life much easier for the modelers. After all, the sum total of human intelligence is no longer conspiring to prove these models wrong. Forrester thought he could build a computer smarter than the market. That was ridiculous, as the market demonstrated in short order. But with "externalities" the market is generally conspiring to make your model come true. Sneaking your effluents into the common river, externalizing your private pollution, your private costs, is an excellent way to boost your profits, if you can get away with it. Nature is perfectly happy to externalize its problems, too: The elephant dumps his dung on the plain and the fox assuages his hunger in the flesh of the rabbit. Neither market nor nature conspires against the pollution model in the same way as the capitalist does, using his every waking moment to conspire against models that predict the exhaustion of rice or oil.

That seems like a very big difference indeed. So big that inveterate optimist Julian Simon was not prepared to bet with inveterate pessimist Paul Ehrlich about such things. He was wise not to. Yet even here, markets will confound the Nuisance models quite as consistently and predictably as they confound the *Limits* models.

THE LIMITS OF MODELS

The first problem, obvious but routinely overlooked, is that human pollutants are by-products of the market, and for that reason alone cannot be any more predictable, over the longer term, than the market itself.

Analysts in Victorian London might confidently have predicted that if the city kept growing at nineteenth-century rates, all the streets of the city would be a foot deep in horse manure by the 1950s. The city did in fact keep growing, but the trajectory of pollution changed course completely. Forrester and the *Limits* modelers failed because markets keep unpredictably changing what we produce, how we produce it, and how we consume it. What we produce and consume determines what we emit. Pollution is indeed a *market* failure, and the details of how markets will *fail* in the future are no more predictable than the details of how markets will perform, the manner in which they will *not* fail. Malthus was predicting "market failure," too, in the temporal domain rather than the spatial. But as markets evolve, so will their externalities. Of that much we can be certain.

The second problem for the Nuisance modelers is that while human intellect is not conspiring to prove their models wrong, the vast diversity and complexity of nature and planetary geophysics have much the same effect. With all life pouring externalities into its surroundings, with granite and volcano, sun and cosmic space, all dumping upon us, too, it is often horrendously difficult to extract humanity's own contributions from all the background noise, particularly as we descend farther and farther below the visible and noxious, into the ever dimmer recesses of the microcosm.

Plants—the green kind, not electric power plants—breathe out 50 billion metric tons of carbon dioxide a year and absorb back 100 billion in photosynthesis. Soil organisms, panting away as they go about their daily chores, emit 50 billion. About 100 billion metric tons physically diffuse into the atmosphere out of the oceans, and about 104 billion diffuse back in. So much for Mother Nature. Environmentally rapacious humans, by comparison, emit 6 billion metric tons by burning fossil fuels and another

2 billion through deforestation. The guess-timated bottom line: a greenhouse profit of 3 billion metric tons. Without humans, it might be a deficit of 4 billion. But we are dealing here with small differences between large and uncertain numbers.

Some of the large numbers are entirely internal. DDT (banned in 1972) and PCBs (banned in 1977) accumulate in body fat and have been blamed for the slow rise in the incidence of breast cancer in the United States. They can act like weak estrogens in the body, and the more estrogen a woman is exposed to, the greater her risk of breast cancer. But overwhelmingly, her main exposure to estrogen comes from her own ovaries. Women secrete enormous quantities of estrogen, at least by comparison with the amount of pesticide they ingest. Breast cancer rates, for example, are very strongly correlated with a woman's own hormonal history, which is reflected in, or affected by, when she starts to menstruate, has a first child, and enters menopause. Those tremendously strong factors must somehow be peeled out of the data before the statistical sleuth can hope to find the tiny effect that pesticides might have. Berkeley toxicologist Bruce Ames adds the observation that plants, too, contain comparatively huge quantities of naturally occurring estrogens and antiestrogens. The amount of biologically active plant estrogens in a single glass of red wine is one thousand times greater than that of all the environmental chemicals that a person gets from pesticides in a day's food. Beans, carrots, and all the other vegetables provide far more.

The same goes for other effects. Pesticides are also said to promote cancer. But edible plants themselves produce all-natural pesticides of their own to ward off the bacteria, fungi, insects, and animals that would eat them. These chemical biocides in nature's own arsenal of self-defense chalk up "carcinogenic" on every standard test for such things, and we consume several grams of them per day, ten thousand times more than pesticides made by man. Indeed, the pesticides permit the cultivation of crop varieties that contain far lower levels of these "natural pesticides." And man's pesticides, unlike nature's, can be washed off before the food is consumed. Similarly, antioxidant food preservatives inhibit the formation of mutagenic, and probably carcinogenic, products of fat oxidation and thus may actually contribute to decreases in cancer. Again, if the modeler misses a small piece when subtracting out any of the big effects, he can be left with any number of small but wholly spurious ones.

Nature's abundant diversity, which the Soft Greens themselves extol at length in other circumstances, also conspires against the modeler. Biochemical effects on one genus are not very good at predicting biological effects on another. And people themselves vary a lot.

The 1977 ban on PCBs was sold politically as a "moral responsibility to the people." And PCBs had indeed been shown to be very toxic, to certain small mammals. But there was no solid scientific evidence of PCBs harming humans in 1977, and none has emerged since. No serious observer believes any more that even occupational toxins, which involve vastly higher exposures, cause "over 20 percent" of human cancers, as federal regulators once declared. "It seems likely that whoever wrote the [estimates] paper . . . did so for political rather than scientific purposes," concluded two distinguished British epidemiologists, Sir Richard Doll and Richard Peto.

If guinea pig models are not good at predicting people, people models often aren't much better. The claimed links between atmospheric soot and SIDS, for example, emerged from a study that ignored California and New York—in other words, ignored cities like Los Angeles, where pollution is high but infant mortality is low—and took no account of two of the most important risk factors: mother's age and alcohol and drug consumption during pregnancy.* High rates of breast cancer in San Francisco and on Long Island can likewise be accounted for fully by demographic, ethnic, and sociological factors: ethnicity, age, age of first pregnancy, and so forth. For reasons unknown, men in New York City have unusually high sperm counts; when they are used as the benchmark, they can make Wyoming look highly polluted by comparison, if you believe that pollution is the main causal factor here. Abstinence time is by far the most important determinant of sperm counts: less frequent ejaculation means higher counts. Three-day variations affect sperm concentrations between 50 and 70 percent. But no sperm-count study can adequately control for how often men of different eras and social environments ejaculate. Nobody has a clue what impact relaxed sexual norms and the proliferation of pornography may be having on this, the most important factor by far. Over longer time frames, geophysical conditions vary a lot, too. Carbon dioxide levels have indeed risen about 20 percent in the last century, and they were only half as high 50,000 years ago, but

---

*The true cause of SIDS is in fact unknown; alarmists have at other times blamed it on the whooping cough vaccine, and drug companies were sued aggressively until that whole theory collapsed into scientific oblivion. We do know that the incidence of SIDS has fallen dramatically since the warnings against allowing infants to sleep on their stomachs or sides became widespread.

they were almost as high as today 150,000 years ago. These things apparently do go up and down. Global temperatures cycle, too.

Nature also has ways of mitigating many effects of concern to the Nuisance modelers, but again, few are well understood. We are all bathed in a sea of background radiation, all-natural hormones, and nature's own pesticides, just as we are surrounded by hostile viruses, bacteria, and larger parasites and predators. Our bodies must have evolved mechanisms to deal with these assaults, or humans would be extinct. In the geophysical context, it is easy to add feedback factors to the models that completely wipe out adverse effects. With global warming, for example, every trend depends on just how clouds form when extra water vapor is added to the air. If the clouds are thick and puffy, they may have one thermal effect; if they are thin and flat, they most probably have the opposite effect. It all depends on whether overall they reflect more inbound sunlight or retain more outbound heat. The only thing that is clear is that the two effects are precise opposites, and nobody honestly knows which, if either, will be important.

Much of the time, one cannot even say for sure whether externalities have bad effects on nature or good ones. "To grass, beetles, and even farmers," James Lovelock points out in his 1979 book, *Gaia*, "the cow's dung is not pollution but a valued gift. . . . In a sensible world, industrial waste would not be banned but put to good use. The negative, unconstructive response of prohibition by law seems as idiotic as legislating against the emission of dung from cows."[3] Carbon dioxide is a poison to us, but it is food to plants. However unwelcome the facts may be, there is an excellent chance that if global temperatures rise modestly, the planet will just grow very much greener than it is today. Granted, Miami might end up under water, but putting an old city above a new rain forest is not how Soft Greens have traditionally arranged their priorities.

Most of the time, the Soft Green modelers try to have it both ways: Nature is both vulnerable to our poisons and essential for absorbing and detoxifying them. There is some truth to both propositions, but there is also a great deal more subtle and complex detail to them than is commonly recognized. And they can't both be true everywhere, all the time. The same is, of course, also true in reverse. Alcohol is a poisonous externality for bacteria; that is why they excrete it into our wine and beer. We drink it with pleasure, and in moderation it appears to be good for health, both physical and psychological. Indeed, as I noted in my introduction, predicted biological effects quite often flip from bad to good as exposure levels drop. For all we now revile "greenhouse gas," greenhouses them-selves are warm and wet and very green indeed.

In general, the long time frame models suffer from all the defects in the Club of Rome models. Dead reckoning is almost impossibly hard. The Micawber syndrome guarantees that if you miss even the smallest bit of favorable feedback, your model will always drift off toward debtor's prison if you let it run long enough. Remarkably—though little noted—the far-future models almost always fail the one obvious test that can be conducted: They are no good at "predicting" the past. The past is the only place where time machines can get tested. If a machine is supposed to transport you back to Pearl Harbor on December 7, 1941, you know what you are supposed to see. If you don't see it, you should properly doubt what the machine reveals when it purports to transport you for-ward in time, to December 7, 2055. Run backward in time, the global warming models cannot reproduce the trends of the last thirty years, still less the last century, unless internal fudge factors are elaborately adjusted to make the retrospective predictions track the known record. In short, these models are to science what Procrustes is to Morpheus. The bed fits not because it is cleverly built to size, but because it cuts off your feet.

Many of the spatial models fail other, equivalent sanity checks, too. Electrification has increased steadily throughout this century. If magnetic fields from power lines were causing leukemia, rates should have in-creased throughout that period. But they didn't. The statistical models quite routinely turn up some mildly *protective* effects just where the mod-elers were hunting for harms. In one of the most cited studies of the sup-posed harmful effects of magnetic fields, it turned out that people exposed at the very highest levels actually had the lowest cancer rates. Several statistical studies, including the largest and most authoritative one, have found that elevated levels of PCBs and DDT in a woman's body slightly *reduce* breast cancer rates. We are quick to dismiss such re-sults as the statistical flukes they almost certainly are, just as we are quick to dismiss the flaky statistical studies that tell us politically incorrect things, such as that the life expectancy of gay males is forty-three years. That the good-news flukes turn up as often as the bad-news ones tells us that it is very easy to conduct serious and elaborate statistical studies that reveal nothing true at all.

With models so easy to concoct and so very easy to manipulate, we end up with far more computer models indicting "harmful externalities" than there are actual harms. Even worse than economists, Nuisance modelers have predicted thirty-five of the last three environmental recessions. Taken all together, the models have "explained" all human cancer, birth defects, and mortality dozens of times over. They have the planet boiling and freezing, too. They have population exploding and also decimated by famine and disease. They have less rain forest but also more; fewer insects, but also more; more cancer, but also less.

It is, in sum, far easier to frame some sort of "scientific proof" of an externality than it is to nail down real health or environmental impacts. It is quite true that we likely undercharge a lot for real externalities. But we overcharge for a lot of unreal ones, too. And rough, scattershot economic justice will not do. Economic efficiency is not a matter of getting the charges right overall, for "industry" or "pollution" in the aggregate; it is a matter of getting them right in the particular emitters of particular pollutants. We tolerate externalities everywhere for the simple reason that life is impossible without them. How much they actually cost us is a quite separate question. The strong evidence is that they cost us far, far less than the sum total of Nuisance modeling suggests.

MODELERS UNLIMITED

The last reason why so many Nuisance models fail can scarcely be articulated, because it smacks of the ad hominem and deeply offends the modelers themselves. It shouldn't. It merely asserts that they are imperfect economic creatures, too, like everyone else. The modelers earn livings, pursue professional advancement, and raise families. They themselves, in other words, are driven by market forces, even if the things they study are not.

And there is a very healthy and profitable market for people who come up with alarmist pollution models. The market in question is funded by government agencies and public institutions and by the private sector, too. Most of the funding is provided for good motives, by well-intentioned people, but motives and intentions are beside the point. The important economic fact is that the Nuisance modelers sell their wares like everyone else, if not in private markets then in publicly funded ones. With one critical difference: In this particular market it is very difficult to distinguish good wares from bad. The whole point of all the modeling, after all, is to calculate and explain things that are not contained, tracked, or evaluated by ordinary market forces to begin with. The market itself is not going to render any verdict on the quality of such models. Good model and bad will have to be distinguished elsewhere.

But the elsewhere is not much good at distinguishing them. As William C. Clark noted in his 1980 essay, "Witches, Floods, and Wonder Drugs," mature academic sciences know how to deal with speculation, conjecture, incompleteness, and outright error. Not so with what historian Jerome Ravetz called the "immature" sciences. There, "a variety of factors conspire to suppress tentative outlooks and to seize on incompleteness as an excuse for polarization. The result is bad science, leading to unnecessary public alarm, unjustified and ineffective regulations." There is "a breakdown of quality control" within the discipline. The relative absence of established facts or criteria of competence tends to make peer review ineffective. Add the pull of a socially relevant, "public interest" discipline, and there is a real danger that the field will experience "an accretion of cranks and congenital rebels whose reforming zeal is not matched by their scientific skill."

And that is just how things have unfolded in the new, immature shadow market for Soft Green science. At the edge, pollution science will always be very immature science, because it can always be pushed to that edge and over it, because there will always be a more diffuse externality to study and a longer time frame to consider. Among the modelers, the chase after any designated molecular suspect just goes on and on. "What is not a risk with a parts-per-trillion test can always be exposed to a parts-per-billion examination," Clark noted. "If rats cope with the heaviest dose of a chemical that can be soaked into their food and water, you can always gavage them. Or try mice or rabbits. . . . There is no 'stopping rule' which can logically terminate the investigation short of a revelation of guilt."

There are, in short, no limits to the growth of pollution models, either. The models keep multiplying and changing for the same reason as pollution itself, as a by-product of other activities: the pursuit of tenure, personal fame, research-laboratory profit, bureaucratic status, political power. If the capitalist has every incentive to dump his effluents on the unsuspecting public, so does the alarmist concocter of models. William C. Clark's comparison with the social dynamic of witch hunting is inflammatory, but not far off the mark. "Where recognition and grant money both accrue to those making the first, loudest, and most frightening noises, where accusations of corruption, cowardice, or insensitivity are the most likely rewards of the careful skeptic, then the 'great confidence game' portrayed by [science historian Jerome] Ravetz cannot be far off."

Diminishing Returns, and Worse

Small wonder that the modern pursuit of externalities is so contentious. Modelers can cook up models quite as fast as polluters can pollute, but there is no easy way to tell which of the models is any good. In such circumstances, it is very easy for "solutions" to transform an ephemeral bad into a concrete worse.

We know for sure that some government policies do that, even if we can't say for sure which ones. Despite common environmental objectives, France, Germany, Japan, and the United States have made quite different regulatory choices about whether to favor fossil fuel or nuclear in generating electricity. Both choices may be bad, as Soft Greens insist they are, but it is impossible to believe there is no difference at all between them. So one country or another has been quite systematically making things worse.

Energy policy makers in the United States alone have done a sort of St. Vitus's dance through the political process, promoting, suppressing, subsidizing, and banning coal, oil, gas, uranium, and hydroelectric power. And citing thin externalities every time: air pollution, oil embargoes, greenhouse effects, China Syndromes, and salmon runs. At the end of it all, we are generating our electricity using a bit of each, pretty much exactly what we'd have done without all the different policies, except that we are using a good bit more coal, the least green option of all.

In other markets, a great deal of effort has been expended just moving things around. Making paper from virgin wood means growing a steady supply of new trees, which sucks carbon from the air, and polluting more at pulp mills out in the country; these tradeoffs may, on balance, be good or bad. Recycling brings more pollution to the city to collect sorted trash, pollutes more water to remove ink from newsprint, but cuts down fewer trees, which may hardly matter when the trees in question are farmed, much like corn or wheat. Coal itself is yesterday's landfill, so there is a perfectly sound "recycling" argument to be made in favor of burying newsprint and corn stalks in modern landfills, where they are effectively mummified, not composted. Doing so could sequester two billion tons of carbon annually, enough to remove from the air much of the carbon that U.S. fossil fuels put into it.

With recycling, the trade-off is often between material and energy. Soft Greens once supported burning trash as another "recycling" alternative, a way to recapture some of the energy content in it that goes to waste when it's buried. But that's anathema today, because trash burns much less cleanly than gas or even coal. Recycling aluminum cans probably saves some net energy, too, but recycling glass or newsprint almost certainly does not. With diapers, ceramic mugs, and newspapers, collection or clean-up generally does end up using more natural resources than tossing the old and making new ones from scratch.

At some point, the hunting down of externalities always turns back on you like that. Dispersion is never cheap or easy to reverse; if it were we would extract gold from seawater, which contains millions of tons of it. At some point, every hunt for game, berries, oil, gold, or sunlight, every filter for soot, sewage, dioxin, or asbestos becomes not merely futile but counterproductive. More new pollution is created than old pollution is abated. This is why trees shed their leaves in winter and why some large animals hibernate. Sooner, not later, the Lady Macbeth syndrome, the ceaseless washing of hands, does not simply fail to remove the damned spot; it draws new blood. There comes a point where leaving bad enough alone is the best you can do and doing more will only make things worse.

Markets discover the problem of diminishing returns quickly enough. But most pollutants aren't currently pursued in markets, they're pursued outside them. And government authorities have a singular power to do things to counterproductive excess. "Malaria" was the original "bad air," attributed to the miasmic vapors emanating from swamps. The way to clear the air was to drain the swamp, which did in fact help, by cutting down on the mosquitoes. Up to a point this made environmental sense, from the human perspective if not from the mosquitoes'. But long after the point was passed, massive government "reclamation" and water management programs were still at it. The federal authorities that once drained the wetlands must now labor to restore them. The government once paid bounties for destruction of "pests" like wolves, so diligently and for so long, so far in excess of what was needed to protect the rancher, that the wolves disappeared entirely. The diligent efforts of yesterday's government to save us from externalities of nature's creation would rank as serious environmental crimes if undertaken by private parties today.

More often, environmental mandates from the government do harm by simply requiring too much effort in pursuit of too little. The Superfund program certainly generates more new externalities in the digging up of old wastes than it abates. Yesterday's market may well have failed; that hardly guarantees that today's government will succeed.

TAKINGS

Economists don't usually put it in these terms, but government, like nature, creates "negative externalities," too. And although the government's intentions are usually more benign than the tapeworm's, the highest law of the land addresses them more severely. Written as it was by people who had plenty of new natural environment but bitter memories of old forms of government, the Constitution says nothing about preserving sky or forest but does proscribe casual "takings" by public authorities. "Nor shall private property be taken for public use without just compensation," declares the relevant clause in the Fifth Amendment.

And what exactly amounts to a "taking?" With the focus now on the public sector's "externalities," the Hard Greens can find takings almost anywhere: In a federal law to protect wetlands or kangaroo rats or seashore, in a local zoning ordinance, in a building permit that conditions approval of a new mall on the building of a bike path, in just about any new rule that limits your right to build, drain, hunt, harvest, or cut down on your own land. The bike path case went to the Supreme Court in 1994. There had indeed been an unconstitutional "taking" there, the Court concluded.

This turns the intellectual tables completely. Now the Hard Greens advance boldly, under the flag of principle. Now the Soft Greens nitpick the terms of their surrender. The rational economist and the principled green unexpectedly find themselves in a quarrel with Jefferson and Madison. It is a quarrel that Soft econo-greens cannot hope to win.

"Externalities" clearly do flow from the public domain to the private; they certainly do every time government condemns a private forest to build a public highway, which government has done for years. "Just compensation" is paid in cases like that; the Constitution so requires. But if courts ever begin entertaining demands for compensation more widely, green regulation will be paralyzed, because protracted litigation throws so much sand into the gears. If they begin actually awarding compensation, it is green regulation that is going to grind to a halt. A lot of what society spends on green objectives today is politically tolerable only because the spending is kept off the public books and concealed from the average taxpayer. It is tolerated, in short, only because it is much like pollution itself: diffuse, insidious, hard to trace, dumped on the unsuspecting public without honest accounting of costs and benefits. But Hard Greens are quite adept at cranking out elaborate models of their own to establish that toxic regulation lowers property values and endangers shopkeepers struggling for survival in the Darwinian world of shopping malls.

What it comes down to is that the whole concept of "externality" is entirely relative. Nobel-prize-winning economist Ronald Coase demonstrated as much in his 1960 paper, "The Problem of Social Cost." When sparks from a train occasionally set fire to farmers' fields, it is as reasonable to argue that the farmers are imposing costs on the railroad as to argue things the other way around, however much ordinary intuition may resist the idea. Legally, a lot depends on who was there first; hence the hog farmer's traditional, common-law defense against the complaints of later arriving homeowners, that it was they who "came to the nuisance."

The Soft Green can only wonder where his once high principle has gone; he finds himself pleading for a little more tolerance and a little more willingness to share the rough with the smooth. Does not government create good externalities, too? Certainly, replies the landowner, and I already paid just compensation for them in my taxes. And as for sharing rough and smooth, why that is just what we have been saying, too, about the smoke from our job-creating factory.

So, in the end, the legal and economic trenches turn out to be dug with perfect symmetry, and we are set for a complete stalemate. For every pulp mill there is a page in the *Federal Register*, a state directive, a local zoning ordinance. Government "takes" from the governed as inevitably as the governed "take" from the environment. Nothing could possibly be more congenial for all the economists and lawyers who make it their business to chase each other in circles.

Markets fail; nuisances are serious and real; and by all economic logic they should be abated. Government prescriptions fail too; takings are serious and real; and by all constitutional logic, they must be abated. Markets are always cheating on their "little" pollutants; regulators are always cheating on their "little" takings. When individual producers can dump the true costs of production onto the environment, they may create the illusion of prosperity as they impoverish society as a whole. When regulators can dump the true costs of government on the private sector, they may create the illusion of good government, even as they impoverish society as a whole. In markets and in government alike, things don't get better, they get worse, when costs are palmed off surreptitiously on others, whether by smokestack or by sleight of the regulatory hand.

There is a dreadful symmetry to all this, dreadful because it can so readily lead to destructive paralysis, both private and public, in markets and in government, too.

3

Complexity:
Gaia and the Sandpile

———— ◆ ————

You've seen it happen yourself with VCRs, microwave ovens, and word processors. Features multiply far beyond our capacity to use them—to the point of dysfunctional excess. Release 7.0 is invariably slower, harder to use, and less reliable than the good old Release 5.3. Same for bank ATMs, children's toys, and cars with computer-controlled seats. Complicated things tend to break down, fall apart, fail to perform. Your old VCR worked pretty well, come to think of it. The new one is so overloaded with buttons and features it's useless. Like a sandpile piled too high, you've seen the engineering collapse, right there on top of your TV.

To judge from accounts in the popular press, it's the same but worse in jet plane cockpits and nuclear reactors. Hairball software and spaghetti hardware inevitably break down. The systems are so complex no one can analyze all possible paths to failure. Disaster must come, sooner or later. Three Mile Island, Bhopal, and Flight 800 weren't so much accidents as predictable consequences of complexity overload. Piling on more safety systems only increases complexity, making unpredicted breakdown more likely still.

That's how Hard technology looks to Vice President Al Gore, too. He invokes the sandpile metaphor, and a grand theory that comes with it, to conclude his 1992 manifesto, *Earth in the Balance: Ecology and the Human Spirit*. Sandpiles inspired one of the big ideas of modern "complexity theory." You can see it in an egg timer. The sand trickles down to form a pile, the pile grows, its sides grow steeper, until finally they exceed some critical slope, and there is a sudden avalanche. Each additional grain produces almost no visible effect until the last one kicks off a massive slide. Danish complexity theorist Per Bak (a physicist when not pondering complexity) argues that sandpiles reveal the operation of "power laws" that govern most of the real world. Bak's own book is modestly titled *How Nature Works*.

Gore, who made a pilgrimage to see Bak, thinks such models explain pretty much everything. "The sandpile theory—self-organized criticality—is irresistible as a metaphor." For Gore it explains the formation of human identity in childhood: "A personality reaches the critical state . . . a person continues to pile up grains of experience . . . sometimes at midlife, the grains start to stack up . . . vulnerable to a cascade of change." It's the same with ozone holes, the same with pretty much any other form of pollution or environmental degradation. The grains stack up calmly, or so it seems, but invariably build up toward an abrupt—perhaps catastrophic—"cascade."

And catastrophic cascade, according to Gore, is where we're headed. "(I)n practice, we have chosen to escape the Malthusian dilemma by making a set of dangerous bargains with the future worthy of the theatrical legend that haunted the birth of the scientific revolution: Doctor Faustus." "Some of these bargains have already been exposed." They are the bad bargains of destroying biodiversity, spraying pesticides, loading up with fertilizers, depleting fisheries, and damaging the atmosphere in ways that increase levels of ultraviolet radiation. The "widespread modern techniques used to squeeze more food from each season's harvest" are doing so "at the expense of future productivity." The food pile keeps growing higher—for now. But the collapse is coming.

Make my day, whispers the technological sandpile. Al Gore hears it.

BRITTLE TECHNOLOGY

They didn't call it sandpile theory back then, but the sandpile was the crux of the original case for "Soft Energy." It was set out in a hugely influential article, "Energy Strategy: The Road Not Taken," published in 1976.[1] The author, one Amory Lovins, a twenty-nine year old living in London, the British representative of the Friends of the Earth, was uncredentialed in science (a master of arts from Oxford was his only degree) but skilled in rhetoric. He has been quoted extensively ever since.

Lovins argued that "soft," home-grown sources of energy—conservation, biomass, solar, wind, low-head hydro—were far better than the "hard" and sometimes foreign ones: imported oil, coal, and nuclear energy. One reason was that we would soon deplete our energy "capital," but never our renewable "income." The other was that Soft technology is stable, reliable, and safe. Hard isn't.

In due course, when it became plain that the capital was not running out at all fast, the second half of the story became the whole story. Six years later, Lovins would title his follow-up book *Brittle Power*. The distinction between "hard" and "soft," he now argued, "rests not on how much energy is used, but on the technical and sociopolitical *structure* of the energy system, thus focusing our attention on consequent and crucial political differences." Soft technologies, unlike hard ones, are decentralized and therefore individually empowering. They are inherently more reliable. "(T)he sheer complexity of many technical systems can defeat efforts to predict how they can fail." Hard options are too vulnerable to sabotage, too prone to catastrophic failure. "The size, complexity, pattern, and control structure of these electrical machines makes them inherently vulnerable to large-scale failures: a vulnerability which government policies are systematically increasing. The same is true of the technologies which deliver oil, gas, and coal to run our vehicles, buildings, and industries. Our reliance on these delicately poised energy systems has unwittingly put at risk our whole way of life."

Two years later—in 1984—Yale sociologist Charles Perrow raised this thesis into a general theory about the fragility of high technology. In *Normal Accidents*,[2] Perrow argued that nuclear power plants, oil tankers, hydroelectric dams, and the biochemical factories of genetic engineers are inherently and inescapably perilous. The same is true of all complex technological systems that handle fuels, chemicals, or biologically potent material. "We have produced designs so complicated that we cannot anticipate all the possible interactions of the inevitable failures; we add safety devices that are deceived or avoided or defeated by hidden paths in the systems." Adding more "safety features" will only increase complexity and make things worse. "[P]erverse interconnections defeat safety goals. . . . [I]mproving or changing any one component will either be impossible because some others will not cooperate, or inconsequential because some others will be allowed more vigorous expression." With "each new advance in equipment or training, the pressures are to push the system to its limits." Putting radar on ships, for example, encourages higher speed; thus "when two ships have radar, they are even more likely to collide." "The dangerous accidents lie in the system, not in the components." The problems inhere in complexity itself.

For Perrow, as for Soft Greens generally, the technology of the atom is the most intrinsically unstable of all. The avalanche is what Chernobyl was, what Three Mile Island might have been, the radioactive sandpile plunging out of control. "The case for shutting down all nuclear plants in the United States seems to be clear," Perrow declares in 1984. The existing plants are not shut down, but the United States does stop building new ones.

Except in spacecraft. Early on the morning of October 15, 1997, NASA launches a Titan IV rocket to carry the Cassini probe to Saturn. Cassini is powered by 72 pounds of plutonium–238, which Soft Greens have long denounced as the most poisonous substance on Earth. "What we are talking about here—and I use this word advisedly—is a holocaust in the making," declares one Soft Green critic. As it happens, Cassini does not blow up on launch, as *Challenger* did. But Cassini is headed initially for Venus and a gravitational "sling-shot" back toward Earth, where it is due to get a second gravitational boost in a 1999 fly by. Yet a single hit by a micro-asteroid at just the wrong moment might knock out the probe's guidance and control systems and steer the probe directly toward Earth. The "fly by" becomes "fly in." At the worst possible angle of reentry, the plutonium will be vaporized, dispersed around the globe, and find its ways into people's lungs worldwide. NASA, declares another critic, is playing "Russian Roulette with millions of lives, up to 60 percent of Earth's population could end up paying for the mistake with their lives. And that's a conservative estimate."*

BRITTLE ECOLOGY

Nature is complex, too. But Soft Greens change their vocabulary here, to escape the pejorative connotations of "complexity." They speak instead of differentiation, genetic variation, the web of life, the intricate interdependence of life in the biosphere, the vast lodes of genetic diversity in rain forests and tidal zones. And of this complexity, Soft Greens speak favorably. Eradicate complexity in technology. But preserve it in nature.

In ordinary circumstances, the Soft Greens say, green complexity is stable and resilient. Perfected over the millennia and left to its own de-

---

*The fly by was in fact completed uneventfully on August 18, 1999, as this book was in press.

vices, ecological complexity is durable, sustainable, and robust. Nature feeds, grows, regenerates, and renews. Lovins says we can live off this biological "income" forever. The wind, sun, and biomass won't be depleted, so long as we harvest lightly. It's the same, presumably, with the cures for cancer and new genetic strains of crops that we will gather some day from the rain forest. Just gather gently and all will be well.

It gets even better. Think of the Earth itself as "a kind of living organism, something able to regulate its climate and composition so as always to be comfortable for the organisms that inhabit it." This is the modest proposal of British chemist James Lovelock, set out in his 1979 book, *Gaia: A New Look at Life on Earth*. Lovelock names this organism after the Greek goddess of the Earth. The "infant Gaia" was born into an atmosphere not friendly to life. Gradually, she "learned" the art of controlling her environment. She developed means of sensing the temperature and atmospheric composition; she contrived ways to control "the equilibrium between the silicate rocks and carbonate rocks of the ocean floor and of the Earth's crust"; she worked out how to fine-tune "the cycling of water from the oceans through the atmosphere to the land surfaces." Individual species—ancient anaerobic bacteria, for example—sacrificed themselves for the benefit of higher forms of life. Life does not merely adapt to the Earth's environment, life controls the global environment itself. The living and the inorganic parts of the planet, air, oceans, and land surface form a complex system that can be seen as a single organism, with its own inherent "capacity to keep our planet a fit place for life."

What reason is there to believe in Gaia? Lovelock observes the amazing resilience and stability of the biosphere, of life itself. The Second Law of Thermodynamics makes such well-ordered molecular complexity "highly improbable." When one does find such stable, complex order, "it is probably life or one of its products, and if we find such a distribution to be global in extent then perhaps we are seeing something of Gaia, the largest living creature on earth."

"But hold on, Pangloss," the Soft Green interrupts. "Let's not let our optimism spin out of control." If metaphors and models are sold à la carte, the Soft Green will take only a small helping of Gaia, along with a large of sandpile.

To the Soft, the interconnectedness story seems about right. Lovelock emphasizes the "tight coupling between the organisms and their material environment," together with "the interactions between the living and the inorganic parts of the planet." "Regardless of the eventual validity of the idea that life controls its environment for its own benefit," declares the *Encyclopaedia Britannica*, Lovelock's "recognition that the Earth's physical, chemical, and biological components interact and mutually alter their collective destiny, by accident or design, is a profound insight."

But Lovelock pursues his metaphor where it leads. And it leads places that do not conform to Soft Green scripture. It leads toward stability, not avalanche. Human activities may disrupt the atmosphere, for example, but then again they may not. Other organs of Gaia—the algae of the sea and of the soil surface, together with the anaerobic microflora in the great mud zones of the continental shelves, sea bottom, marshes, and wetlands—are vigilant, too. Climates are inherently unstable, Lovelock argues: It is life itself that stabilizes them. The planet's entire geo-biological complexity is inherently stable. "The most important property of Gaia is the tendency to keep constant conditions for all terrestrial life. Provided that we have not seriously interfered with her state of homeostasis, this tendency should be as predominant now as it was before our arrival on the scene."

Granted, Lovelock leaves some room here for sandpile catastrophists, but this is hardly a sufficiently alarming model of things for a Soft Green demanding urgent measures to save the environment. Avalanche scenarios are what you need. Al Gore does not emphasize, as Lovelock does, that "our species with its technology is simply an inevitable part of the natural scene." The Soft needs some fragility. Ecosystems are complex, yes: so complex, so interconnected, that they are highly vulnerable to clumsy human intervention. Destroy one critical link in the chain, sever one essential strand of the web, and an entire ecosystem may collapse.

High technology meets nature directly in the context of agriculture, and it is here that the two Soft Green maxims of complexity are applied side by side. Man's complex biochemical concoctions—fertilizers, pesticides, and genetically engineered crops—are rife with peril because they so ruthlessly suppress the complexity of nature. Burn down a rain forest to plant bananas, and you certainly simplify things, ecologically speaking. But the biological simplicity created by plowing and pesticides is dangerously fragile. Allow human eco-engineers to replace a complex prairie with endless acres of simple wheat and prepare for massive invasions of pests, then erosion, then dust bowls. Monoculture farming is brittle and vulnerable because it depends on mile after mile of meticulously bred, single-strain wheat, planted in phosphate fertilizers and doused with pesticides. Stability is to be found only in nature's pest-resistant strains and fertilizers.

And here again, the technology of the atom is the most unstable, the most perilous. Perrow ranks genetic engineering labs alongside nuclear power plants, oil tankers, and chemical factories. And well he might. DNA is an active, reactive, and unpredictably mutable chemical; indeed, one so powerful it can grow itself legs or wings to get around with. In 1983, Soft Green Jeremy Rifkin publishes *Algeny*, a polemical book denouncing all forms of genetic engineering conducted under human auspices. Everything about genetic diversity that is good in the canopies of the rain forest is bad in the laboratory beakers of Monsanto and Genentech.

In other Soft Green visions of disaster, it is the human body that does the amplifying. Instead of the micro-asteroid, a single errant photon of radiation knocks free a single critical atom in a single critical strand of DNA; a cell becomes cancerous and in due course an entire human body collapses. Other models postulate disruption of human immune systems or hormones. A micro-asteroid worth of chemical, the tiniest trace of DDT or PCB perhaps, is said to disrupt these biological guidance and control systems, and 60 percent of the cells of the body, it is said, are put in jeopardy as a result. Once again, the tiniest disturbance manages to set off the massive avalanche.

SANDPILES UNLIMITED

So the sandpiles surround us. Technology is brittle. Nature is brittle, at least when man meddles with it. The more the engineers design and scheme, the more the farmers cultivate and breed, the closer we edge to critical malfunction and catastrophic collapse. It's Malthus, it's *The Limits*, all over again, but with the corners sharpened. The sandpile—or the granary or oil tanker, if you prefer—can grow only so high, and then for one reason or another it will crash. Malthus himself allowed for catastrophe, too. If famine and disease didn't take care of population, war undoubtedly would.

It is in fact remarkably easy to set up computer models that exhibit sandpile properties. The algebra is really quite trivial; it can be programmed in a few lines of code, and the sandpile (or earthquake or seizure) will appear on your screen, rising smoothly then crashing, again and again.

One standard model posits an extremely sensitive but powerful amplifier: DNA, for example, with its self-replicating power to transform injury to a molecule into cancer that consumes a whole body. Climate models assume that carbon dioxide triggers a first little bit of warming. On its own, the effects are inconsequential. But water vapor may play a crucial role in a loop of positive feedback. Warmer air holds more vapor, which blankets the planet a bit more, which warms the air still more, which holds more vapor, and now the Earth becomes a runaway greenhouse. Any model that contains a positive feedback loop will exhibit a propensity for avalanche. This is what creates the howl when a microphone is placed too close to a loudspeaker.

The other standard sandpile model posits a small and delicate control system that controls something much more powerful behind it. Cassini's guidance system, for example. Or hormones, or immune systems. A tiny, unpredictable disturbance (the impact of a single micro-asteroid) is amplified to catastrophic scale (60 percent of the Earth's population in jeopardy) because the aim and timing are thrown off in so exquisitely unfortunate a way. The single untoward grain of sand produces an avalanche of consequence.

Sketch out one or another of these two basic models and the rest is detail. With technology and nature alike, either will readily demonstrate how a simple cascade of small breakdowns might culminate in sudden and wholly unexpected disaster. With high technology, sandpile visions transform the superficially clean into the deep-down dirty. However clean the nuclear plant has been until now, it will be environmentally filthy in the end, after the collapse finally comes. The least-polluting technology day to day may still be the worst decade to decade. No matter how careful the conduct, or how long it has endured, it is the last grain that is going to destroy everything, and you never know how close to the last you may already be.

The calmer things look on the surface, the more dangerous they can be beneath it. If food supplies are abundant, growing even, then the arrival of famine will be all the more abrupt and violent. If the birds sing decade after decade, then spring will fall silent all the more suddenly. The pile of uranium and graphite in the bowels of a reactor looks perfectly stable and secure, until one day it is Chernobyl. The collapse of the sandpile called Chernobyl triggers the collapse of the sandpile called the Ukraine. For Al Gore, the cascade ends up in the polity, mind, and human spirit. Everything collapses but the metaphor, which can apparently rise to any height at all.

The sandpile perspective on ecology and high technology supplies a grand, unifying theory of complexity, and it is important to recognize it for what it claims to be. The vice president of the United States invokes it to sum up his book on the environment. There are plenty of other reasons to favor one technology over another, just as there are plenty of other reasons to protect seashores and rain forests. But in a debate this large and this important, grand theory matters. Especially when it leads to policies that are so very wrongheaded.

SIMPLICITY AND THE SWAMP

Where grand theory leads is to Walden, although not the way Thoreau liked it. If complexity is the problem, then simplicity must be the solution. Low tech, not high. Soft, not hard. "Our life is frittered away by detail," Henry David Thoreau wrote in 1854. "Simplify, simplify."

The first objective is to stop complexity from making any further inroads. It is not easy to shut down a factory or a nuclear power plant once it is fully up and running. It is a lot easier to block the next one from ever getting built. So the main battle lines against complexity are defensive.

Most of the modern apparatus of green regulation has been designed accordingly. By and large, the old is innocent until proved guilty, while the new is guilty until proven innocent. Under the Toxic Substances Control Act, a company must obtain EPA approval before manufacturing or marketing a new chemical. By contrast, tens of thousands of "old" chemicals—listed in an official register—are regulated only if the agency itself takes the initiative. The same goes for pesticides. Old coal-fired power plants have grandfather privileges; new ones are subject to stringent environmental controls. Eliminating old technology is hard; we are reluctant exorcists. Guarding the door against new ones is relatively easy; we have become strict gatekeepers. The inertia of government bureaucracy weighs in favor of the old and against the new.

Burdens of proof are stacked everywhere against technological change and in favor of, at best, incremental improvement in the status quo. When the Occupational Safety and Health Administration (OSHA) wants to ban benzene, an established and widely used chemical, from the workplace, it must prove that there is a "significant risk." But if the EPA wants to approve a new, bio-engineered pesticide, the burdens are reversed. Here, ignorance about risks translates into the strictest possible regulation by the agency. If change is to occur at all, the gatekeeper at the agency insists that it occur incrementally, within the safe boundaries of existing paradigms. Development is preferred over invention. The small step by an established company is favored over a great leap in an altogether new industry. Adding scrubbers to familiar coal-burning plants is easy; switching to cleaner nuclear fuel is difficult.

Nature's complexity is a different matter. Here the mandate is to maintain complexity, not suppress it. Nature's complex interconnections aren't "perverse," as Perrow insists technology's are, they are beautiful. A farmer in a rain forest, like a klutz in a nuclear power plant, can easily set off a chain reaction to disaster.

Yet nothing in Perrow's theory depends on the human klutz. To the contrary, Perrow adamantly rejects the idea that accidents can be blamed on ordinary human error or can be much affected by human intervention of any kind. All highly complex, interlinked systems—nuclear, chemical, and biological—are inherently unstable, however wise or well intentioned their designers. Yes, *biological* systems, too. Perrow himself, as I have noted, ranks DNA right up alongside plutonium.

And this is worrisome indeed, because there is a lot of DNA out there. Nature itself has been mixing and brewing this toxic agent for several billion years, with nothing but natural selection to direct its evolution. By Perrow's logic, the intricate, interlinked, unfathomably complex soup of DNA currently brewing on the surface of our planet should, at the very least, be far more unstable and dangerous than the much more modest cocktails simmering in the labs of Genentech. If nature's brew has simmered longer, that has only given it more time to get more complex, and so, by Perrow's logic, more dangerous.

By Perrow's logic, then, there is only one prudent thing to do: Shut nature down. At the very least, shut down the fetid swamp. Perrow is certainly firm enough in prescribing that cure for man-made complexity like nuclear reactors. But every word he writes, the whole logical structure of his argument, can be applied equally well to the rain forest; only the names of the poisons are different. The unmanaged complexity of the rain forest periodically releases HIV and Ebola viruses, typhoid, cholera, malaria, and leprosy. So drain the swamp, tame the floodplain, burn down the forest, shoot the cougar, eradicate the mosquito, cultivate the field. Let the ancient instabilities of plague and pestilence, feast and famine give way to calm, well-ordered simplicity. The logic is Perrow's, chapter and verse. If such ideas seem outrageous today, they didn't to our grandparents. They did all of that, and more, quite deliberately, and with the best of intentions.

To be sure, a Florida swamp has nothing at all in common with a nuclear reactor. Nothing, that is, except complexity. If complexity obeys any basic rules, if it has fundamental attributes, they must be consistent. Apparently they aren't.

BUICK AND NUKE

Adding complexity does sometimes increase the risk of failure, yet much of the time it seems to have just the opposite effect, with high technology at least. There's no arguing that chips, gadgets, cars, power plants—the whole lot—keep getting more complex, more layered, more interconnected, more of all that Perrow and Lovins say must make things more brittle. But they are not getting more brittle; that is pretty certain, too. If they were, catastrophic accidents would get more numerous year by year. But they don't. All the overall statistics confirm as much: Big, catastrophic failures of dams, power plants, jumbo jets, and chemical factories are growing less frequent, not more so. The catastrophic collapse is becoming less frequent, not more. Most technology does not seem to be governed by the law of the sandpile.

How is that possible? Perrow makes the basic argument that when you pack more bits, pieces, and connections into a system there are more things that can break, more pathways to failure. He's right about that. So, Perrow concludes, failure itself must be more likely. He's wrong about that.

It comes down to Achilles' paradox, one of Zeno's four paradoxes described by Aristotle in *Physics*. Achilles is to run a foot-race against a tortoise, but the tortoise starts out a plethron (a hundred feet) ahead. How can Achilles ever catch up? By the time he covers the first plethron, the tortoise will have advanced another foot. In the time Achilles takes to cover that foot, the tortoise advances again. And so on, ad infinitum. Achilles must catch up an infinite number of times, which must surely take forever. The tortoise can never be overtaken. Perrow's whole argument is Zeno dressed up for a technological age. Practically speaking, the number of possible paths to catastrophe is infinitely large. Humans lack time, resources, and intelligence to analyze or defend against every last one of them. So high-tech risks will always stay out ahead of their pursuers.

The only real problem with this line of argument is that Achilles *does* in fact overtake the tortoise. Observation tells us as much, and mathematics readily shows why. An infinite number of steps can be traversed in a finite amount of time, if the size of each succeeding step shrinks fast enough. Infinite series of numbers often converge to something quite finite. The infinite series $1 + 1/2 + 1/3 + 1/4 \ldots$ does not converge. But the infinite series $1 + 1/2 + 1/4 + 1/8 \ldots$ does converge, to 2. The Achilles paradox simply teaches us an important lesson about the mathematical theory of limits.

And so too it is with the fragilities of complex technology. An infinite (or indefinitely large) number of possible paths to failure can still add up to a perfectly finite risk, and even to an infinitesimally small one. The human body contains billions of cells, any one of which can turn cancerous under an almost unlimited, and certainly unknowable, range of possible assaults. Yet the risks of cancer itself are finite, comparatively small, wholly ascertainable, and in significant degree controllable. A pathway to disaster is only one dimension of risk. The other is its probability. It does not matter how fast the pathways multiply if the associated probabilities shrink faster. Aggregate risk depends on multiplying out pathways and probabilities and then summing them up. Perrow just misses that elementary statistical point completely.

Without walking through all the pathways—which would be impossibly hard in any event—we know that modern technology has them pretty well in hand. Jumbo jets are more reliable than the Wright Brothers' Flyer. O-rings do not fail catastrophically on a tricycle, but use a trike to travel as many miles as a space shuttle and you'll likely encounter a simple but equally lethal failure of some other kind. In car engines and jet cockpits alike, new complexity has generally turned out to be more functional, durable, reliable, and safe than old simplicity.

And despite all the talk of sandpiles and China Syndromes, the nukes of U.S. design keep *not* collapsing. "The case for shutting down all nuclear plants in the United States seems to be clear," Perrow wrote in 1984. "There will be more system accidents; according to my analysis, there have to be. One or more will include a release of radioactive substances to the environment in quantities sufficient to kill many people, irradiate others, and poison some acres of land." There was indeed such an accident two years later. But in the Ukraine, not the United States. Chernobyl proved that a reactor can be designed and operated all wrong. No big surprise there, even if the Russian graphite reactors are quite a lot *simpler* in design and operation than U.S. light-water reactors. But Perrow was quite clear that the accident would involve U.S. operators and design. "There is no organizational structure that we would or should tolerate that could prevent it. None of our existing reactors has a design capable of preventing system accidents." Fourteen years have passed; there has been no U.S. Chernobyl.

"Not yet," Perrow would surely reply. One can always say that about whatever is said to be a sandpile. And people have indeed said it, so vociferously about nuclear technology that we stopped building new civilian reactors. We burned a lot more coal instead: about 40 quadrillion BTUs worth of extra coal since 1989. Nuclear plants, well designed and carefully operated, would surely have done far less environmental harm. We would have done better to fear complexity less and worry about simple old pollution a lot more. Perrow says all nukes are fundamentally alike: too complex to trust. But in complexity, nothing is ever all the same. A swamp is not a nuke, and a Russian graphite nuke is not a U.S. light-water nuke.

Complexity, it turns out, can hide good news as easily as it hides bad. It is true that what we don't know may kill us. But what we don't know may save our lives, too. The only thing that "not knowing" definitely does mean is that we don't know. Perrow weaves a whole book out of the assumption that surprises are always unhappy ones. He is sure there is *one* thing highly predictable about the unpredictable: regret. He is wrong. Alarms and containment systems can perform unexpectedly badly, and sometimes do. They can also perform unexpectedly well, and often do. Jet planes are never supposed to lose all their hydraulics, but sometimes do. Resilient pilots are never supposed to be able to fly such crippled jets, but sometimes do. A nuclear reactor of U.S. design was never supposed to lose all its coolant and melt down, but one did. Its concrete containment structure wasn't expected to withstand the heat of meltdown, but it did. There is simply no law, no engineering principle, no rule of thumb, that requires surprises to be bad ones. There is no scientific link between complexity and catastrophe. If pundits, headline writers, or our own selective memories tell us otherwise, they tell us wrong.

To the extent one dares generalize about technology at all, complexity correlates quite well with better, safer, and cleaner. More gates in a Pentium, more lines of code in WordPerfect, more parts in a Buick, mean more opportunities for a breakdown, but also more opportunities to compensate for failure. There is no empirical basis at all to conclude that software, reactors, or aircraft must be unreliable, dangerous, or dirty just because they're hard to analyze. Making things more complex certainly need not make them better. But making things better very often means making them more complex.

Why should this surprise us? The ultimate good we are pursuing here—nature's diversity—is immeasurably more complex than anything yet designed by the mind of man.

MT. ST. HELENS AND CU CHI

So how about the swamp? Is it stable or brittle? Complexity can never be categorized accurately in any such simple terms. But all in all, there is better reason to believe in Gaia than in Gore.

Nature, in its fecundity, seems to have extraordinary power to recover and recapture landscapes leveled by nature itself and most dramatically by nature's own avalanches. A gigantic nuclear catastrophe blew the top off Mt. St. Helens in Washington in 1980. (The power behind a volcano is, of course, continuous radioactive decay that keeps the core of the Earth hot.) At Mt. St. Helens, the most astounding thing afterward was how quickly the environment recovered. Human attempts to reseed and restore the land had little impact and probably did some harm. A much superior orchestration of the recovery was conducted by nature's invisible hand, the ordinary processes of dispersal, migration, and Darwinian selection. The sandpile did indeed collapse there, on a scale that made nuclear weapons look puny. Nature reversed the biological consequences of the avalanche quickly enough. Given time, the Earth will rebuild the mountain, too, or another one much like it.

Nature often recovers remarkably fast from the worst we can throw at it, as well. In early 1992 I drove through Cu Chi province in Southern Vietnam. The countryside was a rich green; rice fields stretched as far as the eye could see. The entrances to the labyrinthine tunnels in which the Viet Cong hid throughout the war are shaded by lush trees and undergrowth. I was unable to stray from the tourist paths in Cu Chi; the undergrowth was impassably thick. Yet two decades earlier, this area had been barren, as close a thing to a desert as human power could make it. The largest stronghold of guerrillas in the South, Cu Chi had been bombed relentlessly by American and South Vietnamese forces. Defoliants like Agent Orange had been sprayed everywhere. This was one of the few "free fire" zones in the country, where any pilot returning from a mission with unused ordnance could unload before landing. The "American War Crimes" museum in Ho Chi Minh City (Saigon), filled with crude, mendacious, anti-American propaganda, dedicates an entire room (as crude and mendacious as all the others) to the environmental devastation in Cu Chi during the war.

The biological sandpile, in short, was flattened as flat as the technology of that day could flatten it, yet twenty years later it was back up again. Nature, fragile as a lotus blossom, had recovered. Alaska's Prince William Sound has fully recovered from the *Exxon Valdez* spill—and from the even more damaging multi-billion dollar steam-cleaning of rocks, which stripped away the organic seeds of rebirth along with the oil.

No important scientific principle can be proved by examples like these. But there are certainly adequate reasons to put more faith in Gaia than in Gore. One side has the Dodo bird, the other Cu Chi. Some species on the brink of extinction come back, some don't. Forests decimated (it was thought) by acid rain, recover. Others are leveled by loggers and never return. Much of the time, the sandpile doesn't collapse at all, and when it does, it often recovers surprisingly fast. We can indeed bomb, burn, or oil-spill it into submission, for a while, but given half a chance it comes back. This does not sound like a sandpile at all. If it sounds like anything, it sounds more like a sandpile that somehow contrives to fall *up*.

The Limits of Models

Lots of complex, interconnected things aren't sandpiles. When honey descends drop by drop onto a plate, guess what? The collapse never comes, it all stays slow and smooth, and you can stop the gooey spread whenever you like. Gases and liquids hardly ever collapse; their closest equivalents are tsunamis and shock waves. Lots of things aren't brittle, they are resilient, plastic, or viscous. They are regular, orderly, predictable, even in the ways they fall apart, if they fall apart at all. If you believe Lovelock, the biosphere itself is honey, not sand.

It is, in any event, as easy to picture honey as sand, and people have been doing it just as long. "I consider the Earth to be a superorganism, and its proper study is by physiology," declared James Hutton, the father of geology, in a 1785 lecture. Soon after, Thomas Malthus proposed the first and simplest of the ecological "power laws." Herbert Simon, an eventual Nobel prize winner in economics, noted the prevalence of power laws in 1955.[3] Three years later, Arthur Redfield published a Gaia-like paper hypothesizing that the chemical composition of the atmosphere and oceans was biologically stabilized and controlled.

Unvarnished by reliable numbers, "scenarios" are completely useless. It is always possible to imagine something tiny kicking off something terribly big. Micro-asteroid plus Cassini kills billions. Capitalize and add suitable punctuation, and you have headlines for tabloids. Tiny Disturbance Brings Down Nuke! Butterfly Collapses Hoover Dam! Human Genome Corrupted by Microwave Oven! Rain Forest Ebola Threatens Human Race! The "worst conceivable accident" in a bathtub kills every person on the planet. People line up and drown in it, one by one.

But it is just as easy to sketch sandpile scenarios in which the luck is fantastically good. Cassini collides—at just the right moment—with a big asteroid, not a little one, an asteroid that would otherwise have obliterated New York. Cassini's Plutonium Falls in Rain Forest and Serendipitously Destroys Ebola Before It Emerges to Kill Millions. Why not base policy on these hypothetical headlines instead?

Assume enough positive feedback, and things will always blow up; the only question is when. Today's climate models assume water vapor plays a critical role in a positive feedback loop: Warmer air holds more vapor, which warms things still more. If that is the essence of your model, the only remaining question is how soon the thermal avalanche will arrive. It is on the basis of such models that Vice President Al Gore confidently declares in 1992: "Scientists conclude—*almost unanimously*—that global warming is real and the time to act is now." But the climate models of the 1960s focused instead on the cooling impact of smog. Colder air means larger polar ice caps, which reflect more sunlight, which cools things still more. The only remaining question then was how quickly the *cooling* avalanche would come. So *Newsweek* declares in 1975: "Meteorologists disagree about the cause and extent of the cooling trend. . . . But they are *almost unanimous* in the view that the trend will reduce agricultural productivity for the rest of the century" (emphasis added). It's the same with models that assume that biochemicals in our bodies (DNA, hormones, the immune system) will pathologically amplify external insults from radiation, pesticides, or industrial chemicals. Assume a sandpile, predict an avalanche.

But in the real world, there is more stabilizing, negative feedback than destabilizing positive. Real systems generally adapt, heal, and compensate. Lovelock's whole thesis is that the planet does, and the empirical evidence, such as it is, supports Gaia more than it supports Gore. In the global warming/cooling scenarios, big, fat, fluffy clouds have one impact, thin flat ones the opposite one. Neither Lovelock nor Gore can predict cloud shapes with any confidence, but Lovelock can point to a billion years of reasonably hospitable climate. Organisms heal, too. A single photon may indeed knock a single atom out of a critical strand of DNA. But our bodies have been bathed in low-level background radiation for as long as we have lived on Earth, so our cells have probably evolved means to repair DNA, in much the same way as they have evolved means to repair skin punctured by thorns or mosquitoes. It seems unlikely that our hormonal and immune systems are easily disrupted. They have evolved over the ages to exactly the opposite end.

The feedback loops of high technology are designed layer upon layer to be healing and robust. The axiom of safety engineering is "fail operational, fail operational, fail safe": a triple-layering of functionality, so that the first two tiers of failure leave a system still running, while the third-tier failure is at least contained in ways that don't blow things immediately to kingdom come. Yes, *Challenger* blew up anyway. But the actual empirical history of high technology is that we get progressively better at this sort of thing, not progressively worse.

The truth is, sandpiles are too inherently complex for theories about them to yield any practical green advice at all. Most of the time, there is simply no way to know in advance whether a nuke, a swamp, or a Buick has the behavioral characteristics of a sandpile or glob of honey. We can confidently label something a "sandpile" only after we've seen the collapse. Experience—not theory—is what tells us that the sandpile under San Francisco goes critical in a big way about once a century, while the one under San Antonio doesn't. Experience—not theory—reveals propensities of epileptic seizure or heart attack, earthquake or tsunami, tornado or hurricane. Avalanches are not the universal law of nature, they are the universal law of sand and snow. We don't even know if the solar system is honey or sandpile, whether the planetary orbits are stable forever, or whether all the gravitational perturbations and interactions will someday send the Earth hurtling toward Jupiter. We hope for lots of Gaia-like stability here; we hope, like the ancients, for basic harmony and order. But even our best computers can't prove it.

And what the most ambitious sandpile modelers always lack most of all is the feedback of science itself. The big-picture sandpile scenarios all lie squarely in the realm of Alvin Weinberg's "trans-science." They address phenomena that are epistemologically "scientific" but that are simply too large, diffuse, rare, or long term to be resolved by scientific means. Elaborate mechanistic, statistical, and epidemiological models can be cobbled together, but their critical features and assumptions lie beyond verification. At the margins, and here and there, some of the models solidify, some aspects are tested, some are vindicated by actual experience. But for most, the time frames are too long, the effects too diffuse, the confounding variables too numerous.

Soft Greens themselves have said so. The doomsday environmental headlines in Weinberg's day concerned nuclear power plants. Computer models demonstrated that meltdowns were very unlikely. But the Softs scoffed: Such programs could never be trusted. The software shoe is on the other foot now; the machines that said nukes wouldn't melt now say ice caps will. The coal industry scoffs in turn, and with better reason; a planet is a whole lot harder to model than a reactor.

Sandpile science comes down to this: We know how to attach the sandpile label after we've seen a system collapse, not before. And even when we've seen the first collapse happen, we don't know when the next will come. The best we can say of the future is that the frequency of avalanches, small and large, is likely to be the same in the next week, century, or millennium, as it was in the last.

And that helps Lovelock a lot more than it helps Gore. Lovelock infers the Earth has evolved toward stability because it is, in fact, so stable.* Al Gore invokes the sandpile to prove just the opposite: that yesterday's stability is no guarantee of tomorrow's. Gore discerns a sandpile that has been brought by man to the brink of its very first collapse. The whole point of the sandpile metaphor is to scare us out of our complacency. But there is no reliable science, no science of any kind, behind his vision. It is pure metaphor, pushed to silly excess.

Oracles have always claimed to detect something deep in the shallows of human ignorance. The motion of comets and planets across the sky once seemed mystifyingly complex, too. Savants of an earlier age discerned in these tangled trajectories portents of great change in human affairs. Then Copernicus drew the right picture and simply extinguished astrology, the ancient pseudo-science of heavenly complexity. Thereafter, astronomical gyrations in the night sky were understood to portend nothing at all.

Lovins, Perrow, and countless other Soft Green complexity mavens are the astrologers of our own times. Like the astrologers of old, they believe that the complexity firmament really cares how people behave. Baleful complexity will punish us, they predict. What should one make of their enduringly solemn, scientific-sounding theories? Nothing. Complexity is just another name for human ignorance. It augurs nothing, good or bad. Neither engineering nor biological science correlates complexity with fragility, rigidity, resilience, safety, stability, or peril. Complexity has no inherent attributes, none that science can pin down. That's what complexity means: too snarled for human minds to untangle.

So doesn't that very fact—the ignorance itself—counsel us not to mess with it? No. Escaping complexity altogether is not an option. We dwell in a complex biosphere. Starting where we do, the only question is where

---

*Lovelock is sensibly modest about his views. "This book is not for hard scientists," he warns at the beginning of *Gaia*. "If they read it in spite of my warning they will find it either too radical or not scientifically correct."

to wrestle with the complexity that surrounds us, in hard technology or soft, in wood stoves or uranium reactors, in gene-altered tomatoes or in genetic stock culled from the canopy of the rain forest. Fearing the complex nuke, we burned an extra 40 quadrillion BTUs of coal instead, and now we fear the complex greenhouse. The neo-Thoreauvians insist we should neither add complexity to our machines nor subtract it from nature. But it is at least as reasonable to invoke ignorance as the basis for doing just the opposite. As optimistic, problem-solving people have in fact done, with much success, throughout civilization's history.

4

Efficiency:
The Fat of the Land

———— ◆ ————

The *Titanic* weighed 46,000 tons. It burned 825 tons of a coal a day, in 159 furnaces, to heat 29 boilers, which powered its huge reciprocating steam engines. Smoke from the coal poured out from three of its proudly raked funnels. The fourth (closest to the stern) was a dummy, added for grandeur alone. One surmises that the Astors and the Guggenheims had a taste for profligate excess.

Don't we all. In close, most of us are rotten environmentalists. We eat, drink, and smoke the wrong things, and in the wrong amounts, have sex with the wrong lovers, then dose ourselves with the wrong medicines. We breathe, suck, chew, swallow, and inject: incense, perfume, tobacco smoke, silicone, and a rich variety of sexual emissions. We are, perhaps, no longer quite as proud of all this as the Astors and Guggenheims once were. But we do it anyway. The Softs weep. As they see it, it all comes down to fat: fat families, fat cars, fat refrigerators, fat toilet cisterns, and yes, fat diets. We take in too much at one end, emit too much at the other, as we make our elephantine progress across the Serengeti plane of life.

Thomas Samaras, author of *The Truth About Your Height*, pushes the argument to its perfectly logical limit.[1] People are growing steadily taller with each generation. A 20 percent increase in height implies a 73 percent increase in weight. Big people eat and drink more, drive bigger cars, use more fabric and leather for clothes, more gold for rings, and more soap to wash. People are growing taller around the world because,

Malthus notwithstanding, they have learned to supply themselves with a calorie-rich, Western-style diet. Samaras advises feeding children less— a healthy diet but not one that entails "excessive" growth.

We didn't used to call such seemingly private matters "environmental," but as the Softs have been saying all along, the walls between public and private spaces are permeable. Factory, power plant, and ocean liner consume and emit; so do man, woman, and child. Individuals affect the environment by their habits and their numbers. Malthus said populations grow geometrically, and "Malthus was right" declares a man who has proved it himself, Al Gore, father of four.*

To be green we must be pure and thin. Eat organic, not synthetic. Dress in cotton, not plastic. Find fertilizers in the barnyard, not the oil refinery. Split wood, not atoms. Green is pure. Green is efficient. Conserve water, food, and energy. Set things environmentally right close to home and the good effects will surely trickle up the supply chain, trickle up from refrigerator to power plant to minehead, from gas tank to refinery to oil well, from compost heap to the Kansas cornfield, from Sunday newspaper to old-growth forest.

Except that they won't. Not with food, wood pulp, oil, or anything. Purity is not green. Efficiency is not green. However attractive or enriching they may be, purity and efficiency don't directly advance green objectives at all, at least not as prescribed by the Softs. Most of the time, they do just the opposite.

PURITY

Let us start by following our Soft friends on a delightful excursion, to a farmer's market in Paris. The produce is piled fresh and high on rough wooden tables. It has come straight from fields fertilized the old-fashioned way. We pick and choose, only the best, all pesticide-free, the farmer assures us. He wraps it in yesterday's newspaper and we slip it in our string bag. The cheese is unpasteurized; it tastes better that way. The smell of fresh bread drifts from the open-hearth ovens of a nearby bakery. There are meat and poultry stalls nearby, too, and we know the livestock was raised in a pasture, not a factory. And there's not a shred of plastic or preservative in sight.

---

*I owe this crack to P. J. O'Rourke, *All the Trouble in the World* (New York: Atlantic Monthly Press, 1994), 25.

Insist on unadulterated ingredients, go for variety, purity, taste, and the diet will take care of itself. Any Frenchman will tell us that. And though he is completely wrong about everything else, he is right about the most important thing. Getting our food this way is delightful. Immeasurably more so than a concoction of olestra, aspartame, and antioxidant, pre-sliced and packed by machine in a plastic bag.

On such matters, the Soft Green sees eye to eye with our epicurean Frenchman. All these choices are not only healthier, they are greener, too. More frugal. More efficient. And cleaner, of course, by definition. The packaging, fertilizers, and food factories poison the landscape, just as the chemical concoctions poison the body. The genetically altered strawberries most likely corrupt both, displacing natural varieties in the fields and disrupting human biochemistry when we eat them. The whole green revolution, chemically and genetically manipulative as it is, corrupts land and body alike.

So much for food, our oldest form of energy. How about all the others? Gasoline, coal, oil, natural gas, and uranium? Or, if you prefer, solar and wind energy, biomass conversion, and cogeneration. Soft Greens *do* prefer the latter, of course, "soft" sources of energy over the "hard." This was the second, seductive strand of the Soft Energy prescription set out by Amory Lovins in his 1976 article in *Foreign Affairs*.

For Lovins, big electric power plants are the hardest of the hard. About one thousand operate in the United States today, and the environmental case against them seems too obvious to require much in the way of elaboration. They are the environmental equivalent of the factory farm and the chemical-doused field, a green disaster. The huge coal-fired power plant belches smoke, produces millions of tons of sludge, and burns fuel at such copious rates it must be fed by train convoys hundreds of cars long. Huge dams for hydroelectric plants flood vast expanses of land and disrupt the flow of great rivers. Nuclear plants risk meltdown, and nobody wants their spent fuel. "Massive electrification," the entire hydro-nuclear-coal-electric grid, is dirty and dangerous.

Uneconomical, too. Such massive systems entail tremendous waste. The nuke uses the white heat of uranium fission, the technology of the atom bomb, to brown toast a hundred miles away. Using flame temperatures of thousands of degrees, or nuclear temperatures in the millions, for such purposes is "like cutting butter with a chainsaw," declares Lovins. And in between the nuke and the toast, high-voltage power lines slice across the landscape, dissipating power mile by mile.

A policy of relying on such fuels is one of "strength through exhaustion," argues Lovins. It can't be sustained, and it shouldn't be. Nuclear power should be abandoned immediately. Synthetic fuels such as coal gas shouldn't be developed. Other fossil fuels should be used only to complete a reasonably short period of transition. To what? "Soft" forms of energy. To "natural," "organic," "free range" fuel, one might say: Solar and wind energy, biomass conversion (producing liquid fuels from farm and forestry wastes), and cogeneration. Why not warm our homes once again with wood, the simple fuel of old, in a wrought iron stove, the simple furnace that warmed our grandparents so adequately and so well? Lovins's Soft Path will vary from place to place: It may be wind power in Southern California and a small dam in rural Kentucky. The Soft Path uses renewable sources, the hard, depletable ones. The Soft Path spends "energy income," the hard dissipates "energy capital." The Soft Path takes "advantage of the free distribution of most natural energy flows." The Hard is profligately wasteful in hauling energy from place to place. Soft technologies, Lovins argues, are diverse, flexible, resilient, sustainable, and benign. Nuclear power is the hardest of the hard. It is impractical always, everywhere.

A great number of people believed Lovins, perhaps more in Britain, where Lovins spent his intellectually formative years, than anywhere else. Many still do. But the 1970s, the timorous decade of oil embargo and political malaise, were the triumphant years for thinking of this kind. As one 1998 retrospective observed, "[w]e were eager to burn anything back then: wood, methane from garbage dumps, bacterial mats, sunflower oil, buffalo gourd, peanut shells and countless other 'natural' substances."[2] A British company made detailed plans for an electric plant to be built in Denmark, fueled mainly by chicken droppings. An American company undertook the design of a plant to convert sewage into oil. New Zealanders planned to extract tallow from slaughtered lambs and use it as car fuel. A Nobel-prize-winning chemist persuaded Congress to subsidize plantations of guayule, a hydrocarbon rich plant that he was certain would substitute for oil. Norway and India built machines to capture energy from wave power; Britain finally launched a commercial wave-power station off the Scottish coast in 1995. Elaborate plans were formed to build huge floating pistons to capture the energy of the tides and a seagoing generator to extract energy from the thermal gradient between warm surface water and cold bottom water. "Wind farms," heavily subsidized by the government, were built in California, Hawaii, and elsewhere. Sunlight was going to be used to split water molecules in catalytic cells into hydrogen and oxygen; the hydrogen could be used as fuel for power plants or even for vehicles.

EFFICIENCY

Purity is half the Soft Green battle; efficiency is the other half. How much more efficient it must be to eat soft. How much less energy and material it must take to live simply, naturally, and close to the earth. How much more efficient it must be to grow crops without the vast, costly, ruinous, destructive excesses of fertilizer and pesticide.

Efficiency figures even more prominently on the Soft agenda for energy. Most important for Lovins, more important than all the organic fuel, is conservation. "Technical fixes," Lovins insists, can cut energy "waste" in half. "Negawatts" are cheaper, safer, better in every respect than Megawatts. Making cars more efficient is soft, drilling for more offshore oil is hard. Roof insulation is soft, arctic gas, hard. Even better than solar in Arizona or a windmill in California is utility-sponsored home weatherization in New York.

The best thing of all about efficiency is that it entails no pain. Accused of peddling a policy of painful privation, Lovins responds that he "explicitly assume(s) no significant change in where we live, how we live or how we run our society," and that he "goes to a hell of a lot of trouble to make the phrasing accurate."

It is in the promotion of efficiency that Soft energy pundits claim to have achieved the most, the fastest. The drive for efficiency succeeded beyond all expectation. Our ceilings today are insulated twice as well, our walls 40 percent better, our floors four times as well. New furnaces, air-conditioning units, heat pumps, refrigerators, water heaters, washers, and dishwashers all use much less energy than their predecessors. Gas furnaces are 20 percent more efficient, mainly because pilot lights have been replaced with electronic starters. The efficiency of refrigerators has more than doubled; washing machines and dishwashers are 60 percent more efficient. Cars averaged 13.5 miles per gallon in 1975, 22.5 mpg in 1995. Extremely efficient fluorescent lights are proliferating. And almost all of these excellent numbers continue to rise steadily.

In other words, billions upon billions of barrels and watts have been saved by technology that simply made them unnecessary. The Softs quote these statistics all the time. It is easy to convert such numbers into equivalent numbers of oil tankers unfilled and power plants unbuilt. The Softs often do.

Come to think of it, we made comparable improvements in the efficiency of our diets during this same period, too. We zealously "conserved" calories. We favored "efficient" foods, foods that deliver extra miles of repletion on fewer gallons of calories: low-fat milk, diet sodas, and fat-free potato chips. Between 1970 and 1990 the average American has added the sugar equivalent of about a pound a year of artificial sweetener to his diet. We recently added the marvelous olestra to our larder of caloric efficiency. It has the "mouth feel" of pure oil yet is indigestible by the human gut.

Yes indeed, we have certainly grown very smart at conserving calories. Yet our contumacious scale refuses to acknowledge the fact. Could it be broken? We know in our hearts that it isn't. Wardrobes full of clothes that are now several sizes too small tell us the same. The scale is not broken. Despite all those calories conserved, we have just grown fatter.

The Softs were wrong. Completely, laughably, ridiculously, preposterously wrong. For what it's worth, Hard is far more efficient than Soft. But it's not worth what the Softs say it's worth, for the simple reason that "efficient" has almost nothing at all to do with "frugal." This is true for food, and it is true for energy. The whole gigantic myth to the contrary is no more or less than a case study in wishful, credulous, anti-scientific propaganda.

THE BIONIC COW

To begin with, "purity" and "efficiency" are quite different—and for the most part, antagonistic—ends. Efficient things are not pure. Nature, which is "pure" by definition, doesn't care a fig for efficiency, and it isn't efficient at all. Why should it be? The workings of nature were not designed for the satisfaction of some Soft Green efficiency maven in the sky. A mouse has good reason to forage efficiently, but it has no reason to make itself efficiently digestible to a fox or hawk. A tree has excellent reason to transform sunlight and carbon dioxide efficiently into wood and oxygen; a fungus has excellent reason to transform the wood right back into water and carbon dioxide, efficiently, too. This is all about as efficient as two huge diesel engines guzzling fuel and going nowhere in the tractor pull contest at a county fair in Kansas.

Impurity—technological, chemical, biochemical, and genetic—is what gets the engines moving more efficiently our way. Consider the case of a familiar ruminant quadruped. "The cow is of the bovine ilk," Ogden Nash fondly observed, "One end is moo, the other milk." If we think about the cow any more deeply than that, we probably think of it as a natural, greenish sort of beast, and its farm as clean, in an earthy sort of way. Yet as an engine for converting grass to our kind of food, our cow starts out dreadfully inefficient and not clean in the least. Her several ends emit not only moo and milk, but also fats and cowpats, marble and methane.

Soft Greens will certainly agree with that. There are far too many ends because there are far too many cows, the Soft Green declares. Pound for pound, livestock outweighs humans four to one in the United States. Worldwide, there are almost as many domesticated cattle, sheep, goats, buffalo, and camels (an estimated 3.3 billion) as there are humans. Outside of Gary Larson cartoons the cows don't drive around in the family car, but they are powerful polluters nonetheless.

Indelicate though the subject may be, cows are major, unregulated emitters of methane. As a greenhouse gas, methane is much worse than carbon dioxide; cows, it turns out, are thus significant contributors to whatever global warming problems we may face. At one point in a clean air debate, the Senate went so far as to consider methane control options. Scholars pondered the relative advantages of corks (to be placed, with suitable filters, in cows' rear ends), collection bags, catalytic converters, and a brand new gas tax. Moreover, millions of acres of forest are cleared for cattle grazing every year, and cattle contribute significantly to topsoil erosion and water pollution from agricultural runoff.

What the cow emits in other ways, sometimes by way of daily liquid contribution, sometimes by way of the ultimate sacrifice, is fat. As study after dreary study confirms, we should all be consuming less of it, whether it issues from the udder or the flank. Fewer cows would be healthier for us and greener for the planet, too.

Bovine anabolic steroids (growth hormones) do for cows pretty much what human equivalents do for Olympic sprinters: stimulate faster, leaner growth. Bovine somatotropin (BST), the first major agricultural product of biotechnology, can boost milk production by 25 percent. But these chemicals violate the first principle of Soft Green: They are not Pure. We will not find them in the French farmer's market. Some trace of what goes into a cow comes out and might conceivably be harmful. Even if it isn't, we have too much milk anyway, so who needs more efficient cows? Terminally trendy Ben and Jerry's has denounced BST on its ice cream cartons.

There was a day when the farmer's market was about as green and healthy as you could get. When the United States passed its first federal pure food law in 1906, purity and safety were pretty much the same thing. Beyond salt, sugar, and vinegar, most forms of food "adulteration" were bad, almost by definition. That is no longer true today, not even on the health side of the picture.

So far as health goes, eating two all-beef patties ground from a cow that grows lean and mean is much healthier (though less savory) than downing a quarter-pound of bovine Mae West. The health benefits of eating less fat far outweigh any imaginable risks that hormone residues in the meat might cause, especially since the best scientific opinion holds that these minuscule residues are harmless. Use of artificial pesticides permits the cultivation of crop varieties that contain lower levels of carcinogens than those produced with "natural pesticides": those produced by plants themselves to ward off the bacteria, fungi, insects, and animals that would eat them.

The environmental case for the impure cow is clearer still. Growth hormones make cows substantially more efficient at converting grass to human grub. In other words, you can produce the same amount of milk with perhaps 20 percent fewer cows and a concomitant reduction in methane emissions. Gorging on Ben and Jerry's fat-clogged but BST-free ice cream means tying up as cow pasture some land that might otherwise revert to beautiful woods.

Surely, then, the world can use a more efficient, better-tuned cow. One that spends less time ruminating and belching, one that produces more milk for the methane, more meat for the marble. This is precisely what growth-enhancing products provide. We favor more efficient cars, why not more efficient cows? We have the technology at hand to run SUV cows with econo-box emissions. Hard Greens are all in favor of using it.

Hard agriculture is invariably more efficient than Soft. Livestock allowed to wander freely on the open range impose a far greater burden on nature than livestock raised in a pen and fattened on corn. Plucking salmon from a river does far more environmental harm than raising them in cages anchored in a small patch of ocean. As Dennis Avery has pointed out, "Today's typical environmentalist worries about how many spiders and pigweeds survive in an acre of monoculture corn without giving environmental credit for the millions of organisms thriving on the 2 acres that didn't have to be plowed because we tripled crop yields." The bionic cow requires less pasture.

Preservatives and packaging have much the same impact because they reduce spoilage. Hauling produce in the back of the farmer's pick-up, then inviting the epicure to pick out only the best from the stall, makes for the best cuisine and the most wasteful, too. Poor packaging that allows spoilage is as wasteful as a dripping faucet or a leaking gas tank. Irradiating chicken, meat, and spices kills salmonella and extends shelf life. Because almost all food is transported great distances, cutting spoilage in any of these ways is as environmentally useful as improving the gas mileage of the truck. Either way, less gas delivers more food to the intended consumers, to people rather than bacteria and fungi.

Or, to put it another way, fertilizers, pesticides, packaging, and preservatives are the chemical keys to the best solar-power systems in widespread use today. They permit us to capture far more energy from the sun, far more efficiently, using less land. In return for the modest amounts of oil required to make them, they substantially boost the performance of the best solar-power engines yet invented, green plants.

End to end, Hard agriculture transforms earth, sun, chicken, and cow into edible calories much more efficiently than Soft alternatives. Does it also impel us to eat less? Of course it doesn't. "Efficient," as I argue further a few pages hence, is not synonymous with parsimonious or frugal. In fact, they have nothing to do with each other at all.

CUTTING BUTTER WITH A CHAINSAW

Difficult though it may be to believe, Hard is far more efficient than Soft in meeting the rest of our energy requirements, too.

It is, indeed, difficult to believe. Big central power plants seem so obviously dirty and inefficient that Softs almost never bother even to check the numbers. The Sierra Club repeatedly urges the nation to "stabilize national electricity demand," and fancy charities give big grants to promote that end. When General Motors announced its first plans to develop an electric car, the Natural Resources Defense Council dismissed the idea with a sour wave at power plants. "We have to generate the electricity somewhere," grumbled a spokesman, "and that would create a great deal more pollution."

Yet the engineering facts are beyond serious dispute. It is more efficient, and a whole lot cleaner, to burn fuel and distribute electricity than to refine fuel and distribute gas or gasoline. This is true whatever fuel you're using, including the ones Lovins favors—wood, trash, agricultural waste, and hydroelectric power. If you're going to use such fuels at all, big and central is much more efficient. Contrary to all small-is-beautiful intuition, it's better to burn fuel in the external-combustion engine of an electric power plant than in the internal-combustion engine of your average car.

A good electric power plant converts about 35 percent of its thermal energy into electricity, and the best ones approach 50 percent. Subsequent losses in the transmission grid are small. The best car engine can convert barely 20 percent of its heat to torque. Transmitting electricity, charging a battery, and running an electric motor entail some energy loss, but refining oil for cars entails even more. Going electric means lower pollution right off the bat.

So if you have to cut grass two hours a month, use an electric rather than a gasoline lawn mower; leaving it to the central power plant to turn the blade will generate less pollution, not more. Cook a two-pound meatloaf in a microwave rather than a gas oven, for the same reason. Dry the paint on a new car with ultraviolet light rather than gas heat. Produce one pound of steel in an electric mill rather than a blast furnace. Or drive nine miles to work in a car powered with electricity rather than gasoline. Each of these choices redirects fuel consumption from a small, wasteful, dirty, end-of-the-line burner to a large, efficient, comparatively clean one. Each will save about a pint of gasoline or the equivalent and thus avoid emission of about two pounds of carbon dioxide. Emissions of nitrogen oxides, unburned hydrocarbons, and carbon monoxide will be reduced by 90 percent or more. Using electrically produced microwaves or ultraviolet light to dry paint, newsprint, or wet clothes can be still more efficient, because the radiation is tuned or aimed to heat just the right thing. The green who prefers real numbers to metaphor clamors for more centrally generated electricity, not less, whatever the fuel used to generate it.

Harder fuels have a second big advantage: They invariably burn cleaner. However lovely it may look in the rich man's fireplace, wood—biomass—is a filthy fuel. Far worse than coal—which is high-grade, deoxidized wood—because biomass hasn't been compressed and dried as coal has and because it is usually burned inefficiently. Natural gas is largely methane, a far worse greenhouse gas than carbon dioxide.* Beyond gas, the rest of the Soft biases end up, for all practical purposes,

---

*Natural gas contains traces of radon, too, so a gas turbine releases more radiation to the atmosphere than a nuke.

pushing things toward coal—not scarce, not sandpile scary, populist in fact, as it employs honest proletarian miners—and is also the dirtiest option among hard fuels.*

Where it really pays to be efficient, where minimizing consumption of materials and maximizing output of energy is essential, you don't find much that is Soft. Aircraft and military ships use fossil fuel, or better still, nuclear. Solar panels are used effectively (though by no means cheaply) on satellites in Earth's orbit, but they operate in a unique environment of intense sunlight and no gravity. On its Cassini probe to Saturn, NASA used 72 lbs of plutonium-238 to produce electricity for the probe's instruments.

Burning uranium is not easy at all; it takes enormous investment in advanced technology to make it happen. But when you finally succeed, you have a system that extracts the most energy from the least amount of raw material. Hard technology is efficient insofar as smart investment in capital substitutes directly for expenditure on fuel. The whole history of the Hard power industry is one of moving from lower- to higher-grade fuel, from higher-bulk fuel to lower, from less capital to more. It takes more capital to extract higher-grade fuel, more to make it burn, more to contain it while it burns. Anyone can gather wood and burn it; man has been doing that successfully for tens of thousands of year. But you need to gather and burn a great deal of it to roast even a single pig. Gathering and burning uranium is very much harder, but a tiny volume of it, prepared just so, can heat and light an entire city. From wood to coal to oil to uranium, the harder the fuel the less natural resource you need, the more capital it requires to burn, and the more efficiently you can transform stored energy into power. The equation is really as simple as that.

As for Lovins's scorn about cutting butter with nuclear chainsaws, he has the thermodynamics exactly backward—He is just plain wrong about the basic physics. High temperature indicates only that a lot of energy has been released fast, in a small space. How efficiently or otherwise that energy gets transformed into what we ultimately want is a completely separate matter. Thermodynamically speaking, however, high temperature is a sign of *low* entropy, the closest heat itself can come to what you are trying to extract from it, which is useful work. Very high temperatures, in other words, are the one sure indication we have of *high* efficiency, not of low!

---

*Dirtiest for the proletarian miners, too, as countless cases of black lung disease attest.

With today's technology at least, "hard" is in fact thin. "Soft" is in fact fat.* That might change; technology does sometimes change such equations. Mainframe computers were once far more efficient than PCs, but they no longer are today. Energy systems might go through a similar transition. Internal combustion engines, for example, might improve to the point where they beat mass transit, and if they do we will be sorry to have invested wastefully and destructively in things like roadbeds for trains. A fully loaded minivan is already a lot more efficient, per passenger, than a half-full train, especially when you take into account all the extra effort and energy wasted getting to and from the train. But for now, at least, bigger is generally more efficient.

The big, central systems are, for the most part, much more responsible in the handling of their pollution, too. An electric power plant is sited miles from urban areas. Its high-temperature, well-maintained combustors burn much cleaner, and they can be sanitized with scrubbers and such. Small internal combustion engines, by contrast, stick close to people, as we find out quickly enough when stuck in traffic behind a bus or truck. And when it's up to millions of individuals to tune the engine and maintain the catalytic converter, millions won't. Finally, the big plants can use much cleaner fuels, if the public will let them. So long as it remains contained, uranium is a zero-emission fuel.

Soft Greens themselves strongly favor mass transit over private cars. Yet buses and trains are big, relatively centralized, and more polluting, until you take into account the fact that they carry many more people, and then they are recognized to be very clean. If they had to pick their favorite transportation system in the west, most greens would probably nominate the French bullet train. It is indeed a marvelous, clean, efficient, well-run system. Electric, too. And France generates most of its electricity with breeder nuclear reactors.

PROFLIGATE EFFICIENCY

We know we should eat less, less fat, less food in general. We know it would be better for us and (thanks to the learned Dr. Samaras) we know

---

*There is only one notable exception. Lovins favors cogeneration, too, in which factories and others burn fuel on their premises to generate both electricity and heat. The numbers there are very attractive: Such arrangements can indeed beat central power plants hands down, because "waste" heat is put to good use. The private sector knows this, too, and cogeneration has been rising steadily.

it would be better for the environment, too. We know, but we gorge anyway. Richard Klein explains why, in his 1996 exegesis *Eat Fat*.[3] Fat tastes good. Fat is not beautiful—Klein doesn't quite persuade on that score, though he tries valiantly—but fat is indeed very comfortable, especially in the making. We really must cut down, we murmur sadly, as we step off the bathroom scale and contemplate a wardrobe of unwearable clothes. For consolation, we head out to dine with our good friend Klein.

But thanks to Lovins we can now take real comfort in knowing that our refrigerator is far more efficient than it used to be. As everything else he has touted has drifted into oblivion, Lovins can still cling to that one, big triumph. Our ceilings, walls, and floors, our furnaces, refrigerators, washers, and cars are all very much more efficient today than they were two decades ago. Nobody can dispute the real gains that have occurred *there*.

No, nobody can. No more than they can dispute the gains made with sweeteners that contain no calories and buttery tasting olestra that contains no digestible fat. We've been consuming more and more of such low-cal and no-cal substitutes for the last twenty years. And the gains are there for all to see. You can read them on your own bathroom scale. Diet sodas in hand, Americans have grown fatter and fatter.*

With diet refrigerators for our diet sodas, we have gained power plants and coal mines, too. The United States consumed 71 quadrillion BTUs of energy† in 1975 and 91 "quads" two decades later, a gain of a quad a year, the arrival of all that wonderfully Soft efficiency notwithstanding. Fossil fuel consumption rose about half a quad a year.†† Coal has displaced some oil, but year by year, the average American consumes more total BTUs of energy, not fewer. More electricity, too. The average American will increase his annual consumption of electricity by about two megawatt hours in the course of the 1990s, a bit more than his increase in the 1980s.

The most striking thing about the overall trends is that if you chart them over time you simply cannot discern the energy conservation movement at all. With only minor dips and blips here and there, the

---

*Michael Fumento lays out all the dispiriting details in his 1997 book, *The Fat of the Land: The Obesity Epidemic and How Overweight Americans Can Help Themselves* (New York: Penguin Group, 1997).

†A quadrillion is 1,000,000,000,000,000. A quad of BTUs comes in 183,000,000 barrels of petroleum, 38,500,000 tons of coal, or 980,000,000,000 cubic feet of natural gas.

††Some of this story is told in a patchy but useful 1997 book by Herbert Inhaber, *Why Energy Conservation Fails* (Westport, Conn.: Quorum Books, 1997).

curves are all quite smooth and steady. They just keep rising, decade by decade. Judging from the curves, Lovins never exhorted us to conserve energy, and we never responded: Nothing ever changed at all. The only thing you do see is that the mix changes, sometimes for the worse. Some oil has been displaced by an older, dirtier, more greenhouse-gas-producing fuel: coal.

Now, you can blame some of the steady rise on population, which increased from 215 million to 263 million in the same twenty-year period. But even the per capita figures are discouraging. Petroleum usage dropped a bit, from 152 to 132 MMBtu, but coal rose sharply, from 59 to 76. Annual per capita usage of all fossil fuels combined dropped modestly, from 303 to 292 MMBtu. But per capita energy consumption rose from 327 MMBtu in 1975 to 346 MMBtu in 1995. The Softs point out that we are doing better than in the baby-boom 1960s. We are, but for reasons that have less to do with the design of your fridge than with the design of your family. They point out that "rates of growth" are slowing. They are. If you gain five pounds a year, every year, your "rate of growth" (as a percent of your current weight) will steadily shrink, too. But your waistband won't.

The overall trends are obvious to anyone but those wholly lost in their Negawatt daydreams. Electric power—the hard kind—is inseparably linked to economic growth. Gross domestic product grew at 3.9 percent a year between 1950 and 1973, and electricity consumption grew fast. With the stagnant 1970s factored in, the economy grew at a more modest 2.6 percent a year between 1973 and 1997, and electricity consumption grew, a bit more slowly, with it. Hard power and economic output march hand in hand. They are joined at the hip. Efficiency doesn't affect them at all.

Why should we be at all surprised that higher efficiency does not translate into lower consumption? Soft Greens should be the least surprised of all. It was *their* Malthus, after all, who predicted that population would expand to consume the resources available. Efficiency expands resources. No Soft Green Malthusian can take the slightest comfort from the per capita numbers. Population has grown. Total consumption has grown. It has grown while—and indeed because—we have grown smarter at using what we have. At the turn of the last century, electrical generators operated at about 1 percent of theoretical limits; today's generators generate fifty times as much electricity from the same amount of fossil fuel. But overall they also burn fifty times *more* fossil fuel today, not fifty times less.

For a before-and-after comparison, consider again the *Titanic*. Two replicas of the ship are now under construction, almost a century after the original sank.* They will look much the same to passengers, but not to engineers. No modern ship operates on a coal-fired reciprocating steam engine: The new *Titanic* will burn diesel. All four funnels will now be dummies. Five diesel generators will occupy about the same space as five of the *Titanic*'s twenty-nine boilers. Below the waterline, the hull will have a different shape, a bulbous protrusion at the prow. The original's design was good for its day, but naval engineers have since discovered that these dolphin-like additions prevent the formation of an energy-robbing bow wave, improving fuel economy by about 4 percent.

No regulator required any of these considerable improvements in efficiency;† profit-maximizing capitalists discovered and implemented them all. But be that as it may, the replicas are, without doubt, considerably more efficient than the original. So what? Scrapping the other 24 boilers, and all 159 furnaces, will free up lots of space, to carry more passengers, more cheaply. The diesel generators will provide electricity for both propulsion and everything else on the ship: lights, air conditioning, hair dryers, and at least twice as many passenger elevators as the three on the original. Even the Astors and Guggenheims didn't enjoy any remotely comparable range of power-hungry electrical amenities. The replicas will be lighter, more efficient, and safer, more everything that Soft Greens endorse. But not more frugal. These improvements are not emblems of modern frugality at all. They are emblems of modern, middle-class bulge.

THE TRICKLE-UP FALLACY AND THE WEALTH EFFECT

However unlikely it may seem, the individual consumer can easily end up consuming more, not less, as all his motors and engines get more effi-

---

*In 1998, the keel for a full-scale replica of the original *Titanic* was laid in Durban, South Africa; the maiden voyage is planned for December 1999.

†What *has* environmental regulation contributed in this context? In the original *Titanic*'s day, garbage and waste, sanitary and bilge water were simply thrown overboard; replicas will need modern sewage treatment plants and garbage compactors. Ship construction and safety regulations have been tightened, too, of course. *Titanic*'s sister ship *Britannic*, under construction at the time of the disaster, couldn't be built to the original plans; its older sister, the *Olympic*, was put into dry dock for a major overhaul that included massive structural changes.

cient. He almost always does. The Soft Green's fallacy here is obvious. It is the trickle-up fallacy. It is the same fallacy that tells you that when you drink a Diet Coke you are dieting. The problem with trickle-up theories is that the trickling often doesn't head where people assume it must. Most of the time, the Diet Coke trickles down onto a large order of fries.

More efficient means much the same as less expensive. Most of the cost of owning a refrigerator lies in running it, and refrigerators do indeed run much more efficiently than they used to. So people can now buy bigger ones, and that is exactly what they do. And better ones: frost-free units with ice makers, which gobble up almost all the energy saved by the super-efficient compressor at the back. What buyers want is the most refrigeration for the money. A diet-soda compressor at the back justifies the extra-large brownie out front. It's the same with planes, cars, and just about everything else. More efficient engines mean cheaper travel, so you travel more. Today's soda cans contain one-fifth the aluminum they did three decades ago, but the number of cans purchased has grown by an even larger multiple. Fat people often insist their metabolic rates play similar tricks to thwart their diets. But whether or not metabolisms are that nasty (they probably aren't), the energy economy certainly is.

The second factor—an even more important one—is the wealth effect. At some point people just don't want any more ice maker or air travel. But people always want more money, and they find ways to spend it. With all that money saved by your gas furnace in the basement, you fly to Aspen for a weekend in the snow. If appetites didn't compensate for "savings," such trade-offs wouldn't happen. But if appetites didn't compensate, diet sodas and olestra would help us shed pounds, too. They don't.

Indeed, Softs themselves, or their close political allies, have been telling us that for years. In their many campaigns to limit or ban such products, they have invariably insisted that "diet" foods do no good at all. The average American has added the sugar equivalent of a pound of no-cal sweetener to his diet every year, and a pound of real sugar every year, too. People subconsciously adjust. The Diet Coke tastes even better with a double-fudge brownie, and that's how most dieters drink it.

With calories and BTUs alike, the story is just the same. Amory Lovins simply peddles the energy equivalent of *Thin Thighs in 30 Days* to an upscale policy market in Washington. The conservation "savings" have gone the same place as all the "savings" from diet sodas and fat-free potato chips: into belly and buttock. Lovins promised "technical fixes" that would entail "no significant change in how we live." So do all peddlers of quick diets. So do the people who sell aspartame and olestra, though they are generally more honest in that they promise less. But the thighs are not getting thinner, not in thirty days, not in thirty years. The fact—the dismal fact—is that we have done what the book told us to do, but the cellulite is still there.

The only surprise is that so many are surprised to hear this. Most Americans want to cut down on calories, but don't, even though the results of failure are personal and unpleasant, even though success is so visibly healthier, sexier, and rewarding in every way. How then can we expect success with the low-cal car or fridge? Most other forms of "fat" are not ugly to the average eye, they are rather attractive, which means that shunning is harder still. Unlike a banana split, a bigger refrigerator won't make you feel miserable the day after; it will just make you think how convenient it really is. You need it for all your diet soda. Big cars do ride better, especially on big highways. As Madison Avenue well knows, a great number of American drivers would rather like to drive 125 miles an hour on an autobahn: The nasty truth is that more than a few people find the idea thrilling, some even apparently consider it sexy. And if the woman in the passenger seat of the Mercedes is blonde, the man in the driver's seat is tall. The five-foot-four Samaras says it himself: Tall men get better jobs, better money, better girls. The psychological deck is perniciously stacked against the small.*

So "efficient" has nothing to do with "frugal" at all. Our atavistic appetites, our genes themselves, remember famine and cold too well to be fooled by Nutrasweet. However much our plenty, we will hunt and gather more. It isn't good for us, and it isn't good for the environment, and nothing stops it, nothing at all, not efficiency, not politics, not even war. Nobody who has actually looked at the scale and examined the long-term trends can possibly believe otherwise.

Or that, in any event, is where I shall leave it for now. As I argue in Chapter 9, efficiency does turn out to have one great, green grace, after all. The right kind of efficiency, the kind discovered by markets rather

---

*Even the fates seem to conspire against efficiency. Samaras doesn't quite finish the story. Big people require larger coffins when they die—more resources, there—but they die earlier. "(A)bout one year per inch" earlier, Samaras notes. What he does not note is that during their extra years of life, the short still eat, drink, drive, dress, and wash. Which— if you crank the numbers—suggests that on a life-time basis consumption by short and tall may pretty much even out, in the end.

than by Soft Green regulatory bureaucrats, makes us richer. And among its many virtues, wealth is green.

COMFORT FOOD, CALORIES, AND QUADS

Lovins prescribed comfort food to a nation snarled in endless lines at the gas pump. His story was perfectly tuned to its political times. Significantly, it was published first in *Foreign Affairs*. Why there? The Arab Oil Embargo had just hit, more threatening to the American way of life, it seemed, than nuclear weapons. America wondered if she might go out not with a bang, but with a whimper, not in fire, but in ice. Arabia, our best-loved McDonald's, had announced it would stop serving up the quadrillions. Lovins proposed some home cooking, from ingredients we already had on the premises, that no Arab could take away. We had plenty of nothing, but nothing would be plenty for us. It would be a safe plenty, too, not a scary one. Within three years Three Mile Island would make nukes scarier still. Nukes offered the energy equivalent of plastics and pesticides. Lovins promised unscary plenty, home grown, natural, free of all poisonous preservatives.

It sounded so plausible, so reassuring, so good. What a pity that it was pure bunkum, beginning to end. Technology has grown steadily more efficient in every decade of this century, including the last two. Efficiency remains a perfectly sensible thing to pursue, in power plants, refrigerators, in agriculture, too. It is as useful to save energy as it is to save time or Christmas wrapping paper. Which is to say, sometimes it's worth the trouble, sometimes it isn't. For the happy few who include them in a real diet rather than a fake one, low-fat potato chips offer a chance to brighten an otherwise spare meal. That's about all the people who sell such chips ever promise. Which makes them the most honest people in all the energy conservation chronicles. Lovins and his acolytes sold something more along the lines of *Thin Thighs*. There was to be no real pain. We could eat what we liked, so long as it was all tofu or pure chocolate washed down with wine, or macrobiotic pounds of avocados.

Catering to people's unscientific tastes and fears is perfectly legitimate. Soft sells. Madison Avenue knows that, too. It invents entire campaigns around fictitious uncles who brew beer on their back porches and fictitious aunts who whip up pancakes on their very own griddles, just before bottling, boxing and delivery to a supermarket near you. Mom's cooking tastes like love, even if it tastes like lard, too. It is perfectly fair to insist that it tastes better than a tofu burger, better than any gene-altered tomato drenched in artificially colored ketchup. It is equally fair, of course, to prefer Mom's station wagon to the school bus. We just aren't free to assert that lard is healthier than sushi, or that cars get better per-passenger mileage than buses. The station wagon certainly feels safer than an econo-box, driving 70 mph is more convenient than driving 55 mph, and safety and environmental effects are what they are, not what we might wish them to be. Calories and quads are not hard or soft, they simply are. Fat and thin are not states of mind, they are states of avoirdupois. And anyone who cares to know the truth about *that* can examine the scale. The needle keeps rising.

Thus, obesity and cancer become, correctly, our new, post-Malthusian metaphors, the metaphoric opposites of starvation. Malthus said we would starve when we ran out of land; the Softs now tell us we will run out of land because we gorge. We swell and swell until we have absorbed everything: the whales, the redwoods, the elephants, the sandworts, and all the rest. The fatter the humans, the thinner the land.

Yet that, curiously, is the one thing that is *not* happening. There is something important we *have* begun to save in recent decades, something that we are using much more efficiently, and using less of, too. What we are using less of is land, the Earth itself. I return to this kind of efficiency, the only one that matters for environmental purposes, in Chapter 7.

5

Eschatology:
From Malthus to Faust

◆

Malthus was green, too. Men of his day had to be: Nature supplied them with all their food and fuel. Malthus simply argued that when mankind reached the limits of nature—when it had farmed all the farmable land—mankind would starve. With the technology of 1798, this was both obvious and true.

Two centuries later, the modern Soft has updated the script only a bit. We need the Earth for food, fuel, and its gene pool, the source of new crops, fuels, and medicines. Nature processes our sewage, sucks carbon dioxide from our air, and detoxifies our countless other excretions and emissions. Nature serves as biochemical sentinel, too: When sandworts pine, frogs die, and ospreys are unable to reproduce they warn us of pollution that endangers us, too. We must regulate air and water pollution, mainly to protect human health. Officially speaking, we banned DDT not because it harmed birds but because it might cause cancer in humans.

Nature as a whole will take its revenge if we abuse her. Paul Ehrlich attributes the AIDS pandemic to the "deterioration of the epidemiological environment which is quite directly related to population size as well as to poverty and environmental deterioration." Global warming means more mosquitoes, concludes a Yale virologist, which means new epidemics of dengue fever and yellow fever in North America. Alterations in the geophysical environment will ricochet, too: The human body is adapted to the pre-industrial atmosphere and ozone layer, the normal background levels of radiation and ultraviolet light in which our paleolithic ancestors evolved.

But are these facts true? Hard Greens doubt that they are. We deny that "carrying capacity" is a quantifiable metric of science. We discern no ineluctable tie at all between nature's decline and humanity's.

The "harvest" case for environmentalism is history. Every year we manage to grow more food on less land. We used to harvest fuel and rubber from the forest, but not today. Nor do we need Brazil's rain forest to supply genes that we can engineer at will in laboratories in New Jersey. Human ingenuity transcends nature's limits. The *Limits* prophecies always fail.

As to swallowing our wastes, nature is indeed good at that. Dumping our sewage into rivers and lakes breeds more new life than it suppresses; the water we like the best is generally the deadest, short on algae, weeds, and all the other greenery that thrive on our excretions. And here too, we readily improve upon nature: Nature consumes our wastes best under factory-like control, with septic systems run much like breweries. Growing new trees is an excellent way to pull carbon from the air. Filling modern landfills is another. We do not maintain Central Park to absorb New York City's wastes; we build dumps and treatment plants.

And if most of nature is an excellent recycling plant for our wastes, which it is, it can't also be a hypersensitive canary, imperiled by the same. Other species do teach us some useful things about human biology, but most low-level biochemical effects are genus-specific. A hormonal protein screams "sex" to the cow, but for a human body it is just lunch. A mouse isn't even a very good model of a guinea pig; a frog is certainly not a good biochemical model of a man. When the EPA administrator declared DDT to be a potential human carcinogen, he had to overrule the findings of his own scientific review committee. The ban was good for birds, but no serious evidence of benefit to human health has ever emerged. Visions of nature taking its revenge by way of mosquitoes, fevers, and such are about as scientific as a Jay Leno monologue. Why should only the species that hate humans thrive in our despoliation?

As for the geophysical environment, our bodies undoubtedly are adapted to the conditions in which our ancestors evolved. But those same ancestors adapted to the Rift Valley and the Himalayas, to sub-Saharan Africa and the Arctic, too, which is to say, to a wide range of background radiation, ultraviolet light, oxygen, and ozone. And here, too, technology improves on nature, readily shielding us from environmental variation and excursion, as it has since man mastered fire. The idea enrages the Softs, but not the Hards: People can indeed wear dark glasses and suntan lotion if they must. They already do.

SURVIVAL OF THE FITTEST

Most fundamentally, the Hard Green refuses to view the whole of natural creation in the same light as a dairy cow. Cows have been bred to wish us well, but nature as a whole has not. Nature does not wish man good or evil, it does not wish him anything. It wholly lacks an attitude.

We know that on the authority of Charles Darwin. The gazelle's genes have no interest at all in the ultimate survival of the cheetah's, and the same goes for the cheetah's, vis-à-vis the gazelle's. The tapeworm evolves to reside in our intestines; our bodies evolve to expel it. Humans have no scientific reason to believe that "the balance of nature" is generally good for them; indeed, they have no reason to believe in ecological "balance" at all. The whole notion of ecological "balance" is anti-Darwinian. Evolution does not progress toward balance; evolution is the flight from it, the consequence of imbalance.

There is, in short, not the slightest reason to believe that, at this precise point in evolutionary history, man happily finds himself situated in the best of all possible ecological worlds. That eco-philosophical edifice—which, from Genesis to Malthus, occupied the center of the environmental faith of our ancestors—collapsed into nothingness on April 2, 1953. Until that day it remained reasonable to think of life as inherently complex and mysterious, Darwin notwithstanding. Then we discerned the double helix. Now life was just a chemical, a fairly simple one at that. Darwin had begun the demystification. Watson and Crick finished the job.[1]

Ironically, Malthusian greens make the mistake of elevating humans too high. By arguing so broadly that by harming nature man will inevitably harm himself on the rebound, they implicitly assert that nature was created for man's own principal benefit. "(W)ithout man," Kant declared, "the whole of creation would be a mere wilderness, a thing in vain, and have not a final end." The Malthusian Green says the same thing, only backward: Without the wilderness, man himself is extinguished. The Soft Green affirms, in other words, a new Creation Science. It is every bit as unscientific as the old.

The Hard Green believes *The Origin of Species*, instead. And the one profound ethical fact that emerges from Darwin is that nature as a whole is neither benign nor malign; it just doesn't care. Science has a phrase for this philosophical opacity, this moral blindness, this complete and absolute indifference to the interest of any and all: Survival of the fittest.

And the trouble for the rest of nature these days is that our species is very fit indeed.

The Ascent of Man

Mankind apprehends today the basic, bio-engineering mechanics of creation. We understand how the blind watchmaker builds the eye itself.[2] We understand the smallpox virus, an ancient, evil scourge of humanity: understand it so well, in fact, that we have eradicated it from the face of the planet while simultaneously conserving it cybernetically, as a complete, paper record of its genetic code. We could recreate it from scratch if we chose. We are now poised to eradicate ancestral perils embedded in our own genes.

Give us another few decades—an inconsequential tick of the clock in the span of biological history—and our species' mastery of the rest of nature will be complete. After the flood, God directs Noah to "subdue" creation, to take "dominion over the fish of the sea, and over the fowl of the air, and over every living thing that moveth upon the earth." Today, we can think of nature as benign only because we have obeyed that one command so very faithfully. We have no more practical reason to conserve nature than we have to conserve cows. We can subdue at will and replenish at will too, with transgenic mice and cloned sheep. The fear of us, the dread of us, is upon every living creature that dwells on Earth.

If we must now conjure up new gods, it is comic to imagine them dwelling inside redwoods or blue whales. Genentech's laboratories are more like it. From viruses to whales, nature's power plants were designed by a blind watchmaker. Vaccines and nuclear power plants were designed by watchmakers who can see, and very well. To be sure, some human watchmakers are inept or evil; some genetic accidents of nature are equally baleful, if blindly so. That only tells us we have to pick and choose among the new techno-priests and prophets, just as we picked and chose among the passion fruits and poison ivies. High technology is complex, but less complex than nature. It is fickle, but less fickle than nature. It is unstable, but less unstable than nature. From pesticides to vaccines to nukes, high technology is an environmental peril not because it often turns against us but because it usually does exactly what we ask of it. It makes us so very fit to survive that nothing else in Darwinian creation really stands a chance.

The new ecological hell now in sight is very different from the old. The Malthusian hell is as black as the waters before God's creation; the Faustian hell is merely beige, the color of man's concrete and computers. In the Malthusian hell, the ascent of man causes the collapse of everything else and that, in turn, destroys man, too. The whales and the ocean drown first, followed almost immediately by all the human occupants of the lifeboat. In the Faustian hell, the ascent of man causes the collapse of everything except man. Everything sinks but humanity.

At this point in history, the second vision is a lot more likely than the first. We can go it alone. We need energy, nothing more, and we know how to get it from many more places than plants do. We don't need the forest for medicine; as often as not, we need medicine to protect us from what emerges by blind chance from the forest. We don't need other forms of life to maintain a breathable balance of gas in the atmosphere or a temperate climate. We don't need redwoods and whales at all, not for ordinary life at least, no more than we need Plato, Beethoven, or the stars in the firmament of heaven. Cut down the last redwood for chopsticks, harpoon the last blue whale for sushi, and the additional mouths fed will nourish additional human brains, which will soon invent ways to replace blubber with olestra and pine with plastic. Humanity can survive just fine in a planet-covering crypt of concrete and computers.

COMPLEXITY UNLIMITED

There is not the slightest scientific reason to suppose that such a world must inevitably collapse under its own weight or that it will be any less stable than the one we now inhabit.

Basic science has only a few, modest things to say about complexity. In one of its several equivalent formulations, the Second Law of Thermodynamics states that the entropy of any closed, self-contained system tends to increase. Orderly, organized forms of energy—a tightly coiled spring, for example—decay into disorderly, chaotic, heat.

At first blush, life itself seems to violate this law. The history of life on Earth is one of more orderly (lower entropy) systems evolving out of disorderly (higher entropy) ones. But this is easily reconciled with the Second Law; the Earth isn't a closed system. The Earth stands in the river of energy that flows from white-hot sun to black-cold outer space. The planet continuously absorbs more ordered, higher-temperature, lower-entropy light from the sun and emits more disordered, higher-entropy heat into the deep cold of the cosmos.* These are the conditions that make the evolution of life thermodynamically possible. Not inevitable, just possible.

One can pursue this line of analysis only a little further, and only to suggest, not to prove. In his 1996 book, *Full House*, Harvard paleontologist Stephen Jay Gould sets out to explain why life on Earth has grown steadily more complex since the dawn of creation.[3] There was no other direction to go, he concludes. Life started simple and stupid, so it had only one way to go from there. As with a drunk who starts leaning against a wall, the stumbling can only head one way, and that's away from the wall, toward the ditch. It's a purely random walk. The only order to it comes from the two boundaries to the problem—wall of simplicity, ditch of complexity—and where the drunk starts: up against the wall of simplicity. If we managed to transplant some higher forms of life to a completely new environment—Mars, say—the evolutionary stumbling would then progress in the other direction, too, back toward simplicity. "The vaunted progress of life is really random motion away from simple beginnings, not directed impetus toward inherently advantageous complexity."

Gould neglects only to emphasize how much life itself can accelerate the stumbling. Life, by definition, manages the flow of energy through its own, local environment. Once a system is able to do *that*, it can begin to march toward complexity, not just stumble. It eats, grows, and replicates, and over time it evolves. Food chains transform the lower complexity of the grass into the higher complexity of the rabbit. Supplies of light, land, and water limit the aggregate complexity of any given habitat, but complex life can alter those supplies. A beaver builds a little dam. Herbert Hoover builds a big one.

So the growth of biological complexity is likely to follow a hockey-stick curve. It may take a billion years to get from single-celled to multi-celled animals, another two billion to get from plants to animals, and yet another billion to get to primates. But primates with fire can become primates with nuclear reactors in a mere fifty thousand years. Complexity has the singular power to create more complexity and at an ever-accelerating pace.

---

*Life is likewise possible around thermal vents deep in the ocean, where geothermal energy flowing from the Earth into the cold water creates thermodynamically comparable conditions.

Nothing in the laws of thermodynamics or in the history of life on Earth proves that this must happen. Only that it can, and that it did.

So far as the basic science of complexity is concerned, Soft Greens like Lovins and Perrow have it exactly backward. It is instability that generates complexity, not the other way around. The instability of the sun, pouring its energy out into the blackness of space, makes thermodynamically possible the redwood, the whale, and man himself. Life, the highest complexity we know, is the product of solar-energized instability, on a large scale, over a long period of time. Instability is the motive force that makes possible the evolution of complexity.

Technological complexity is merely a recent (and still comparatively modest) extension of life's complexity. By all indications, techno-complexity will continue to rise, at accelerating rates, for the indefinite future. Whether or not civilian nuclear power ever rises again in the United States, the technology of atom and photon will prevail. Gene technology is ascendant in both agriculture and medicine. Silicon technology and fiber optic glass are propelling the information revolution. In the networked microcosm complexity doubles and redoubles every year. Information is complexity incarnate.

If one is to project forward at all, the most likely trend is more of the same. Mankind's complex, energy-channeling systems won't collapse into destructive chaos, as Perrow says they inevitably must. To the contrary, each advance is likely to spawn further advance. In the thermodynamic environment we occupy on this Middle Earth, it is possible and reasonably probable, that complexity will give way only to still higher complexity, indefinitely into the future.

In all likelihood, then, high-tech environmental hell is perfectly feasible, sustainable, and viable. Humans multiply as the sands on the shore. The entire surface of the planet ends up like Manhattan, without Central Park. These thoughts are repellent, but that does not make them untrue. Our most likely future is a high-tech hell: comfortable, stable, sustainable, perfected in every way for the comfort of our species and no other.

Unless we consciously choose a different one.

PART II

Capitalist Green

6

The Conservative Commune

———————— ◆ ————————

The instincts ran in the family. In the 1860s, T. R.'s uncle, Robert Roosevelt, had gained international attention for his efforts to save the shad run on the Hudson River. In the 1930s, T. R.'s cousin Franklin would establish national parks and forests, while his Civilian Conservation Corps tended forests and built wildlife shelters. But it was T. R.'s own administration that transformed public attitudes about the wilderness and placed conservation on the national agenda.

Few in today's political climate can easily credit the historical fact that this was accomplished by a conservative Republican president. Until T. R., the federal government's main role was to acquire new wilderness, not infrequently by force, and then induce people to plow and farm it. The Homestead Act moved the prairies from public to private hands. Until T. R., there were no limits to the wilderness, no prospect we would ever run out: at least, none to speak of officially recognized in Washington. The scarcity that the federal government had resolved to eliminate was scarcity of cropland, settlements, cities, lumber mills, canals, and railroads. Animals, habitats, rivers, and forests were the problem. People were the solution. Until T. R.

Roosevelt affirmed the opposite and translated it into ambitious political action. Large tracts of wild space were to remain public property. They should be held in perpetuity by government, all the people together, rather than by private owners. They would be managed, if at all, by public officials, not private interests, with at least one primary objective being to leave them as they were, unspoiled. T.R., it would seem, was a strange political beast. Rough Rider. Conservative Republican. Socialist Green.

CALL OF THE WILD

Today's political stereotypes do not admit to the possibility of such a political centaur: half capitalist, half commune; half man, half nature. Liberals are supposed to be the ones who affirm the value of public ownership, public programs, public space. Conservatives are supposed to shun all things public. They privatize: railroads, highways, hospitals, and schools, and they'd privatize the Grand Canyon if they could. Or so the political agendas of Left and Right are typically portrayed.

What was T. R. himself thinking? As I noted in the introduction, he had learned his conservation the hard way, on his Chimney Butte ranch in Dakota. The public prairies were overgrazed, and everyone was the poorer for it. Upon his return to New York in 1885, T. R. dashed off a book, *Hunting Trips of a Ranchman*. It got a generally enthusiastic, but also somewhat critical, review by the editor of *Forest and Stream* magazine, George Grinnell. Roosevelt immediately went to visit him. From Grinnell, an older man and a more knowledgeable one on environmental matters, T. R. learned that the hunting he so loved had already been reduced to "a thin remnant of what it had been merely a decade before." "With the zeal of the new convert," Roosevelt established the Boone & Crockett club. It was dedicated to the conservation of large game animals and their habitats. Roosevelt's passion for conservation evolved from there. It would weave together three main strands of passion and logic.

As a recreational hunter T. R. knew exactly why he needed to conserve the wilderness. That was where the game lived. Where else would he find a cougar to shoot? Tellingly, T. R. named his Boone & Crockett Club after the two hunters who had helped open up the frontier.

T. R. the rancher contributed a second perspective, one that betrayed a surprising lack of faith in the free market. Private enterprise had no thought for the future, or in any event, not enough thought. Only government husbandry would protect future supplies of grass for livestock, lumber for logging, and water for hydroelectric power. T. R. the rancher believed that the point of conserving was to continue using nature:

forests for lumber, ranges for grazing, rivers for electrical power. Cattle-men and ranchers were to be involved in public conservation and would come to embrace it, because it was in their own private interest to do so.

These first two strands of his thought ranked T. R. as a Hard Green of his day and put him in sharp conflict with its Softs. T. R. was no Ralph Waldo Emerson, Henry David Thoreau, or John Muir. Those "preser-vationists" revered wilderness for its own sake. Muir, founder of the Sierra Club, adamantly opposed building the Tuolumne River Dam in Yosemite to supply water to San Francisco. T. R. supported it, consistent with his "wise use" philosophy of conservation. As president, T. R. "had no desire to close woodlands wholesale to loggers like those bluff fellows he had met while hunting . . . nor did he desire to run his cattlemen friends off the ranges of the West. Such people, he remained convinced, were the bedrock of the republic; their activities ought to be encouraged and their livelihoods protected."

But the third strand of T. R.'s environmental thinking brought him a lot closer to the preservationists than either camp acknowledged. Roo-sevelt was a conservationist for purely aesthetic reasons, too: He simply loved the wild, loved it passionately. It was in 1910 that he wrote the words quoted at the front of this book: "(T)here are no words that can tell the hidden spirit of the wilderness."

CONSERVATIVES AND THE COMMUNE

All in all, T. R.'s instincts have proved remarkably prescient. So prescient, in fact, that to this day both the logic and the passion of his environ-mentalism reflect mainstream American thought on these matters.

The strongest objections come from the political edges of the green debate and focus on the first two tenets of T. R.'s environmentalism. The softest of today's Softs, like the preservationists in T. R.'s day, object to the hunting and the "wise use." Like Thoreau and Muir, they still want nature preserved for its sake, not for ours. The hardest of the Hards in-sist that T. R. was quite mistaken in believing that only government fully grasps the future value of ordinary economic assets like timber or grass-land. They believe—correctly, in my view—that markets can always beat government in managing and conserving resources for strictly economic ends, like the production of lumber or beef.

It is the third strand of T. R.'s thought, the aesthetic one, that has proved the most politically durable. Even a hard-riding Republican conservative, it appears, can discern in the unworn wilderness something that inspires awe, that commands respect, something that is, in the deepest sense of the word, simply beautiful. Soft Democratic liberals undoubtedly can discern it, too. And both still do, in considerable numbers. Indeed, their ranks swell year by year, on both sides of the political divide.

And it is also this third strand that is the most troubling for many conservatives, particularly those of an economic bent. With all ordinary kinds of property, the fence contributes to the value and preserves it. With wilderness, it is the other way around. We consider great, wide-open, public spaces to be valuable—*economically* valuable, if you insist—precisely because they are great and open.

From an economic perspective, in other words, we love them for entirely negative reasons: because they are not contained, not tamed, not productive, not private. The value inheres in the fact that anyone can contemplate them or hike them or simply take pleasure in his tiny—but indefeasible—joint-tenancy in something so vast and open. A whale in Sea World is instructive and entertaining, but it is not the same as a whale whose migrations span the globe. In every way an accountant, economist, or even an ecologist might measure, Disney would operate Yellowstone much better than the National Park Service. But Yellowstone would be diminished nonetheless. A vital part of its grandeur, and our own, is that it belongs not to Wall Street but to America. Value that inheres in citizenship, nation, patriotism: Such values cannot be contained or conserved in any private market. With such things, to privatize is to destroy.

Many conservatives will respond that public ownership often destroys, too. Yellowstone itself has been badly mismanaged.[1] Private interests capture public agencies; the upshot can be public prairies run by the rancher lobby, national forests by big lumber, and continental shelves by big oil. Roosevelt was quite mistaken in his belief that government conservation would benefit not only the land but also the cattle rancher and the lumber company. For many decades, the government stewards of land, rivers, and swamps did nothing but pave, dam, and drain. When government has attempted to balance conservation with strictly economic interests, it invariably manages to despoil the land and dissipate the economic opportunities, too.

Yet even so, conservation may be a mission that government can learn to perform reasonably well. The one thing that big government is capable of doing well is doing nothing, which happens to be the paramount objective of conservation. Private interests can conserve, too, and do it very well, and most conservation initiatives will remain in private hands where they properly belong. But at some point, the vastness of White Mountains and Everglades, of river archipelagos and coral reefs, the sheer scope and scale of conservation at its most ambitious, will require a conservator that is commensurately large.

Most of the time, and happily so, conservation works fine on a much smaller scale. All in all, private conservation is, by a wide margin, the most important form of conservation we have. A great deal of conservation occurs on entirely private ranches and estates, private lands, shores, and lakes. Private land trusts are by far the most important and fast-growing factor in the conservation movement today, particularly in the rural West and Southwest.[2] Originally established in New England, they have recently been embraced by ranchers and mountain communities. There are now some twelve hundred local land trusts, and the number is growing rapidly; they represent an estimated one million members. In the past decade alone they have set aside an additional 2.7 million new acres—an area almost as large as Connecticut—for conservation. For the most part, the trusts do not own the land; they own, instead, conservation easements donated or sold by nearby landowners. In mid-1998, Montana had 258,416 acres under conservation easement, more than any other state.

Nor does public conservation have to be federal; most of it is better managed by local or state governments. Conservation efforts by counties and states surpass the federal government's, as they should. Florida, for example, is spending $300 million a year to buy land for conservation, some six times as much as the National Park Service.

That conservation is often the product of private initiative will not surprise conservatives, and Hard Greens will never call for federal management where private, local, or state initiative will do. But an incontrovertible fact remains: Some values depend on doing things on a scope and scale that is inescapably public. It is through our public acts, after all, that we attend to things very much larger and consequential than ourselves. We little note nor long remember what happened in the private fields and houses of the men who drafted the Constitution, or fought at Gettysburg, or confided Yellowstone to the public trust, but the freedom and beauty they conserved still shape and enrich our lives. Private stakes in things so large can never be large themselves, but they can be very durable, far more so than anything contained in private space.

Indeed, one of the things that most sharply divides the conservative from the modern "liberal" is the liberal's reckless willingness to level what took centuries or longer to grow in public soil, his propensity to pollute religion, poison morals, despoil culture, and upset the delicate balance of our old-growth Constitution. It is ironic, and altogether wrong, that such people should nonetheless represent themselves as anointed conservators of forest, prairie, river, and shore. Conservation is the political heritage of the conservative. We conserve paintings and manuscripts, furniture and buildings, churches and liturgies, forests, rivers, lakes, and shores. We should not be ashamed to conserve, not privately, and not publicly, either. No holy writ of free-market capitalism, no axiom of laissez-faire economics, no conservative theory of politics or government, requires us to do otherwise.

To conserve wilderness legacies and to expand them is not to abandon conservative principle; it is to affirm it. As we grow older and richer, we resolve to conserve more: more law, more history, more freedom, and more public land, too. And so we should, political conservatives especially. This is the most conservative conviction we can embrace.

SEPARATION OF POWERS

The one major lesson we have learned since T. R. is that conservation under government auspices should be just that, no more. The preservationists like Muir were more right than T. R. allowed, although for the wrong reasons.

Government is quite capable of setting aside wilderness and conserving it in that state, for no use other than recreational, which is to say, for no productive economic purpose at all. But in fact the federal government has generally undertaken to manage its lands for both conservation and economic objectives and to pursue them simultaneously. Roosevelt himself thought this made sense. Experience has taught otherwise.

The whole point of conservation is to be anti-economic. The wilderness is, by definition, that which has *not* been subdued, tamed, graded, plowed, or paved for the direct economic benefit of man. Conservation and economic exploitation are—in the short term, at least—a zero-sum game. Cut down a single tree in the forest, and what you gain in lumber you have lost in conservation.

Policies framed in political arenas almost never rise any higher than that. Try as it may to attend to the future, politics is a short-term business. When government holds a forest for both conservation and lumber, how much we get of each and on what timetable can only be determined by never-ending political struggle. Today's political contest must be won today; tomorrow must take care of itself. A government agency established to balance the opposing objectives of economic use and conservation is a political house divided against itself, forever tugged in opposing directions. The agency will staff up with conservation biologists on one side of the building and economists on the other. But how many trees are cut and how many are left standing will, in the end, have nothing to do with either discipline.

It is much easier, and politically far more stable, to designate particular places for conservation alone. Human exploitation ends with recreation, and the human footprint is kept light. The forests are hiked, camped, and, up to a point, hunted, but not lumbered or grazed. This is the kind of focused, internally consistent, and essentially passive objective that even big government can advance. Or so we must hope, for the spaces that are too far flung to be conserved by anyone else.

The essential political corollary is to keep the government's footprint equally light in the private sphere. To begin with, that is the only way to promote effective, private conservation. Lumber companies need trees for the short term and the long. They know that better than the government, and—despite all claims to the contrary—are far better at planning prudently for the longer term than the government ever is. Lumber companies do indeed cut down trees. They grow new ones, too. It is only by letting them do both, according to their own judgment, on private land or in private nature conservancies, that the government can unleash the free market's own, powerful incentives to conserve. The same holds for hunting, fishing, ranching, and all the other forms of "wise use" that Roosevelt foresaw. Given the freedom to manage what they own, private owners and private conservancies will indeed conserve. In the aggregate, they will end up doing most of our conserving, and doing it better than government ever could.

They will exploit, too, sometimes commingling the two activities, sometimes separating them in space or time, in ways far more varied and subtle than central managers can ever grasp. Conservation and economic exploitation are indeed irreconcilable, but only superficially, only in the snapshot of the short term. In the longer term, under sufficiently involved and smart private management, there is no reason why both objectives cannot be advanced. Lumber companies need forests today, and tomorrow, too. They can be trusted to grasp that very much more clearly than a schizophrenic government agency ever will.

This brings us back to one of the most pernicious aspects of the Soft agenda, its contempt for all bounds between public and private space. Your lightbulb, flush toilet, and hair spray, your washing machine and refrigerator, your compost heap and your right to an abortion, are all of legitimate interest to Soft Green authorities. Theodore Roosevelt would have had no quarrel at all with the intentions behind the Endangered Species Act, which passed the Senate unanimously in 1973. But it was written loosely, and enforced in a manner that pays no respect at all to the notion that public interests must somehow be kept separate from private ones. As amended and interpreted, it proscribes "habitat modification" anywhere, for any reason, once a protected tenant appears on the scene. If an endangered tapeworm happened to find refuge in your intestine, it would be a federal crime for you to consume the medicine that rendered your guts uninhabitable to your guest. The same goes for endangered cockroaches under your floorboards, or an endangered strain of typhoid in your water well. There is no private left, not body, floorboard, land, home, or farm, not once an officially endangered species has designated it as its "habitat."

For conservatives, by contrast, conservation happens in well-defined places, places you can see and draw on a map. Yellowstone and Yosemite start here and end there. Bison, eagles, and rivers are only somewhat harder to track. Our kind of conservation is not about the endless, futile, computer-driven pursuit of stray molecules. This is fundamental: We affirm that workable boundaries can be maintained between the public sphere and the private. We categorically reject the thought that conservation requires huge computer models; or a new oligarchy of scientists, regulators, and lawyers; or interminable, officious meddling with our lightbulbs, flush toilets, and hair sprays, washing machines, refrigerators, and trash. It is Yellowstone that is to be walled off, not the ordinary American citizen. In environmental matters, as in others, the public sphere must be kept separate from the private.

But it is difficult—often impossible—to maintain such fences, the Softs will respond. The natural environment itself does not have fences and will not tolerate them. Sooner or later, every wall, every pipeline, dam, path, and road becomes an environmental obstruction. So it does, the Hards must concede, and that is just too bad for the environment. Beavers build dams, too, and in so doing, they destroy one habitat as they create another. Elephants, giraffes, and whales all transform the habitats that they occupy. Humans have habitats as well and, within broad limits, must control them for their own ends. It is only by separating the private habitat from the public one, as cleanly and consistently as possible, that we will promote the best of both. The only effective way to achieve that separation is to separate powers: the government's, over public spaces, and the private sector's, over private ones. Yes, the lines between the public space and private will always be blurring. Eagle and tapeworm recognize none of these same distinctions. But humans do, and must, if they want to establish any durable equilibrium between man's interests and the rest of creation's.

Liberals and libertarians are of course wont to say much the same thing about cocaine, pornography, sodomy, and abortion. They are the first to complain when conservatives suggest that private conduct has public consequences. They argue, cogently, that maintaining good fences between private and public life is essential for both, for civil liberty on the one hand and good government on the other. So it is. But the debate is always how high those fences can be and when they are crossed. Murder in the privacy of the bedroom is still murder; shooting eagles and wolves has environmental consequences even if they are shot when they stray onto private land. Public/private walls will never be as simple or clean as political factions might like, but maintaining them well is nonetheless essential.

In accepting and embracing government's central role in conserving trans-economic public goods, conservatives can be all the more adamant in opposing government meddling in ordinary economic affairs. It is because we want the government to do conservation well that conservatives should insist on shrinking government's role in agriculture, lumber, electric power, canals, railroads, highways, flood control, home mortgages, insurance, and more. The whole point of conservation is to curtail normal economic activity in certain well-designated places. A conservation area is, by definition, a place where normal economic activity is simply prohibited. Charge the government with promoting both wilderness and economic output, and the wilderness will be on the same road to ruin as the economy.

THE CONSERVATIVE ENVIRONMENT

There was never much high-church philosophy to T. R.'s conservatism. It was inspired by aesthetics, an abiding appreciation for the grandeur of nature. And it was disciplined—this may seem a curious thing to say of a man like T. R.—by a real sense of humility. Not much philosophy there, but enough.

Like many a modern curator of the painted word, the modern green is not concerned with aesthetic sensibilities. He is offended—outraged—at the suggestion that a kangaroo rat, not being beautiful or demonstrably useful, may not matter so very much. Beauty has nothing to do with it, he insists, utility still less. Environmentalism is sub-sensual. He is paired off on *Crossfire* with an ugly utilitarian—today usually pictured as a Republican preppie, though a New Dealer or Stalinist from the 1930s could surely have filled the chair equally well—who frames the whole debate as spotted owls versus jobs for loggers, and demands jobs. Beauty has nothing to do with it, he insists, utility is everything. Environmentalism is uneconomic.

They are both wrong, and for much the same reason. It is our capacity for awe, our instinct for reverence, our affinity for beauty, that makes conservation worthwhile and that makes some things worth conserving and others not. The spotted owl should be saved. The smallpox virus should be extinguished. A sense of aesthetics will get you a long way in environmental discourse. It will in fact get you as far as you should go. As we have seen, T. R. had a good one. As biographer H. W. Brands notes, Roosevelt described Africa in language worthy of Homer:

> In these greatest of the world's hunting-grounds there are mountain peaks whose snows are dazzling under the equatorial sun; swamps where the slime oozes and bubbles and festers in the steaming heat; lakes like seas; skies that burn above deserts where the iron desolation is shrouded from view by the wavering mockery of the mirage; vast grassy plains where palms and thorn-trees fringe the dwindling streams; mighty rivers running out of the heart of the continent through the sadness of endless marshes; forests of gorgeous beauty, where death broods in the dark and silent depths.... (T)here are no words that can tell the hidden spirit of the wilderness, that can reveal its mystery, its melancholy, and its charm . . . the strong attraction of the silent places . . . the awful glory of sunrise and sunset in the wide waste spaces of the earth, unworn of man, and changed only by the slow change of the ages through time everlasting.

Seeing things that way inspires, to begin with. Then it liberates. It lets you cut through the scientism, the fussy bureaucratic detail. It lets you ignore the priesthood and dispense with their soaring intellectual cathedrals. It saves you the enormous expense and inconvenience of digging up New Jersey and conserving your own trash. It lets you spend your energy and dollars on places that are awesome, fascinating, or simply beautiful. It lets you oppose the destruction of a forest for no fancier reason than that the destruction is ugly.

The aesthetic approach does not mean ignoring the micro-environment completely, still less rejecting every commandment ever subscribed to by the Soft Green priesthood. Priests, propagandists, and the culture at large help shape our aesthetic preferences, for better or worse. They have every right to paint the environmental word as they please, they just should not palm off their art as science. Purity is beautiful, and industrial by-products in our drinking water are ugly, however invisible and harmless they may be. Fluorine and chlorine in the water are sort of ugly, too, even if they give us healthier teeth and guts. Composting doesn't improve the environment, it quite possibly harms it, but tending the soil beautifies the gardener, just as hunting, however paradoxically, nourishes love of the wild. There is an aesthetic case to be made for frugality. We aren't going to run out of space for dumps, but garbage is not beautiful regardless. Making do with less often is. The Quakers are on to something. It is indeed a gift to be simple.

Profligate excess is as ugly in the digging up of dumps as it is in the original dumping. Roosevelt conservationists will devote far more energy to parks and forests, sewage treatment and cleaner smokestacks, far less to part-per-billion traces on dioxin. We will plant more trees and attend fewer world conferences on global climate. We will cultivate our own garden.

But suppose pesticides really do silence the spring? Suppose the models are right? Some of them certainly are. The trouble is, we don't know which ones, and we aren't going to. There are already too many of them, and there will be an endless number more. In the face of such inundation, plain old conservation—just loading up the ark—is the best practical response. Whatever impact pesticides have, setting aside 100 million acres of forest will likely protect more birds than trying to bankrupt Dow Chemical through Superfund. The most beautiful way to purify water is probably the most effective too: Maintain unspoiled watersheds of the kind T. R. would have been proud to protect. While an "almost unanimous" priesthood forecast cooling in 1975 and warming in 1992, the humble conservationist just planted trees, the most pleasant and practical way to suck carbon out of the air, however little or much that may affect global climate.

Short as it is on crisis and catastrophe, Hard Green rhetoric is less dramatic than Soft, but it is a lot more durable. Hard Greens like T. R. can't peddle their green as essential to the protection of human wealth or health, because they don't believe it is. I suspect many Softs don't either. But they go on peddling the economic and pseudo-scientific rationalizations for biodiversity because they think these are the only ones they can sell to American business or politicians in Brazil. Perhaps they're right. But it's a risky strategy to promise an economic boom that will probably never come or to predict environmental catastrophe that won't materialize either. When the economic case for biodiversity collapses, as it inevitably will, a much more fundamental truth may be lost with it.

And when people find that their own health and material welfare are not in fact threatened by environmental degradation after all, they may stop paying attention entirely. Characterizing problems as very much worse than they are can do much harm in the long term, too. Once it becomes clear that there is no human health effect, the temptation will be to believe there is no environmental effect either. A democratic society pays a real price when accurate information and objective reason are displaced by myth and metaphor.

A sense of humility is needed, too. Roosevelt was irrepressibly confident in his own ways, but never about the ways of nature. Traveling and living in the wilds had taught him diffidence, respect. That is a frame of mind to cultivate in a political milieu so prone to oligarchy.

Environmental awareness gives the Hard Green even more reason, if he needed any, to be wary of grand, public works. T. R. endorsed his share of them; FDR endorsed many more. In retrospect, it seems clear conservatives should have opposed more of the megalithic government projects of those days. They certainly should going forward. Yesterday the federal dollar erected huge dams and drained swamps; today federal money is used to unleash those same rivers and convert sugar plantations back into swamp. The swamp programs are doubly expensive because the government also props up the price of sugar, which makes it far more expensive to return Florida real estate to the alligators. A consistent conservatism might have blocked more of the before and thus saved us from having to do much of the after. Going forward, it should. The more consistently we oppose New Deal government subsidies of development and industry, the stronger our political position will be in opposing new green subsidies of efforts to tear down the same. With all the money saved, we could buy a lot more forest, mountain, and shore.

A consistent philosophy of moderation and caution can also do much to blunt the vindictive, punitive impulses of Soft Green. And that, too, will help make things greener. In the aftermath of the *Exxon Valdez* spill, the multi-billion dollar steam-cleaning of rocks in Prince William Sound did far more harm than good, stripping away the organic seeds of rebirth along with the oil. The Softs' frenzied demands that Exxon be made to pay and pay overwhelmed everything else; increasing the damage to the oil company became much more important than abating damage to the Sound. "Nature, it seems, fared better on its own," *Scientific American* concluded, in places where the cleanup was left to the wind and the waves. The litany of demands for gold-plated remediation comes from people more interested in cutting down capitalism than in growing trees.

And finally on the subject of humility, the Hard Green conservative will seriously engage the micro-environmental Softs, and with an open mind. One does not have to believe in the models to wonder whether there may not be something congenitally reckless in a species so clever, so capable of reshaping the world it inhabits. Are we to become Jonah in reverse, swallowing the whale, the ocean itself? Some changes should be just too big, too fast, too much for the entire comfort and satisfaction of someone who styles himself either "conservative" or "green."

THE MALTHUSIAN CONSERVATIVE

Wilderness conservation expands the common space, in much the same way as conservation of a city park or monument. Garrett Hardin reminds us that the common pastures in England were always overgrazed. The short answer to Hardin is that things like cow pastures—resources intended for economic use—should not be held in common. The "Tragedy of the Commons" comes from collectivizing *economic* resources, not uneconomic ones. The whole point of conservation is to designate uneconomic forests, uneconomic lakes, uneconomic shores, uneconomic wildlife.

The modern conservation movement thus stands for the Malthusian prophecy turned on its head. There is scarcity after all, but not of farmland or food or any of the ordinary economic resources needed to sustain the human population. Man the thinker does not run out of ways to feed his body, except when degenerate political institutions of his own creation contrive to starve him. What grows scarce isn't the land man can tame, but the land he chooses not to.

Enduring material poverty is not inevitable, as Malthus thought; in free markets it is wealth that is inevitable. Free markets are in fact so very powerful at spreading material human wealth so very widely that we now need government programs to institutionalize poverty, at least here and there. Conservation areas are poor, in Malthusian terms: They produce no food for human consumption, they generate no ordinary economic wealth of any kind. We guard them jealously for just that reason, because beauty is scarce, too, and because the wonder of the wilderness is something that no market can create.

7

Save the Earth

——— ◆ ———

England was once heavily forested. The forests were all cut down, first to clear land for farming, then, at a much faster rate, to build warships and provide fuel. They never grew back.

The people of the day saw it coming. As the wealthier elements of society began to grasp what was being lost, forest conservation rose to prominent concern in seventeenth-century England and France. Then England found a substitute—coal. Coal was soon being used widely to make glass, bake bricks, brew beer, and heat homes. Newcastle eventually emerged as London's principal supplier. Coal made possible cement, bricks, and steel, which eventually replaced hardwoods in construction of buildings and ships. Had coal emerged a century or two earlier, the great forests of England would still be standing today.

In this century, too, we have witnessed an even more remarkable environmental shift: the rapid *reforestation* of the American continent. For the first time in history, a Western civilization has halted, and then reversed, the decline of its woodlands. What happened? Quite simply, America reduced, dramatically and rapidly, the amount of land it uses to produce its food and energy. The recent trends in the oceans and waterways under U.S. control are equally positive. One cannot say the same of most other countries. But we can say it of our own.

There is nothing ambiguous about the trends. So far as the simple, most direct, and most important measure of environmental quality is concerned—the total area of the Earth's surface directly exploited for human consumption—the trends are incontestably positive. People are still getting fatter, but no longer off the fat of the land. Not any more. The land itself is getting fatter again, too.

The credit does not belong to T. R. and the conservation movement. Still less to the Soft Greens. We are saving the Earth with the technologies that the Softs most passionately oppose.

LIFE ON THE SURFACE

Before getting into the technology, however, we should pause to savor what an environmental miracle this truly is.

All modern estimates of species extinction assume a one-to-one correlation between acres put into human use—for agriculture, shopping malls, or anything else short of nature conservancy—and the reduction of wildlife in the biosphere. Most of the displacement is occurring in the Third World, for agriculture. Acres of cassava are not acres of rain forest. Cow pasture isn't prairie. Without doubt, the best environmental strategy of all is to literally save "earth": not just land, of course, but also stream, river, lake, and the life-bearing layers of ocean and air.

That is, indeed, the *only* effective strategy. This is just a matter of basic ecology. Life on Earth has evolved on the surfaces and at the edges, at the interface between land and air, land and water, water and air. With few exceptions, all life on Earth dwells within a few hundred feet of the interfaces. The less we disturb the interfaces, the better it is for the biosphere. It is as simple as that.

The Soft Greens say as much too, although rarely so directly and never drawing the right conclusion if they do. Where they say as much is in every one of their sandpile scenarios. With oil, the primary environmental concern is not the derrick or even the offshore platform, still less the drill bit a mile below the surface. The concern is not oil securely contained in the *Exxon Valdez*, it is oil spread for miles across Prince William Sound. It is spill, the ruptured pipeline, or the burning of the fuel, all of which spread the liquid fossils and their by-products across the surface, where the life is. With nuclear fuel, the environmental concerns likewise center on the accidental spreading or on the deliberate spreading by way of terrorists or nuclear war. So long as the fuel remains contained, from extraction to final disposal, nobody can much doubt that nuclear power is environmentally superior to the alternatives.

With all these technologies, in short, the Soft concerns center on the collapse of the sandpile, the accidents that might spread fuel or its by-products far and wide over the surface. So long as the materials in question stay deep underground or secure in the tanker or concrete reactor vessel, the surface, where the life is, stays green. Soft and Hard part company only insofar as they sharply disagree about the stability of the pile. The Softs believe that uranium is safe for the environment only so long as it remains buried wherever geophysical history buried it. Hards believe it is at least as safe when extracted, enriched, and contained in a reactor of human construction. Both believe that these are not materials we want spread across the Adirondacks, the Everglades, or Prince William Sound.

If the Softs are right about the sandpiles, then of course we should not be building them, not if we care about protecting life on the surface. If the oil and uranium we extract from deep in the Earth are bound to end up being spread far and wide across the surface, then we should obviously stop extracting them. If the Softs are right, high technology just provides new chainsaws for leveling the forest, albeit at more unpredictable intervals.

But suppose they are wrong about the sandpile. What then should true greens think of high technology? What then should true greens think of the Softs themselves?

LIFE IN THE THIRD DIMENSION

For most of history, humanity has grown on the Earth's surface alongside all other species, and humanity has gradually subdued them as it grew. For most of human history, we have contrived ways to disturb the Earth's surface more, rather than less. Every advance in human technology was used to extend our dominion over that surface. It could hardly have been otherwise. We evolved on the surface, too, so that was of course where we hunted, gathered, and cultivated. The surface—land, river, and shallow coastal waters—supplied all our food and, until very recently, all our fuel. To this day, our main use of land is to capture solar energy. Not with solar panels or windmills, but by growing crops.

And, until recently, the more our appetites and populations grew, the more land we consumed. Consumed it to produce our food, to supply building materials for furniture, homes, and ships, and especially to produce our energy. Capturing useful quantities of solar energy is extremely land-intensive, for the simple reason that solar energy is very diffuse. The most efficient way we have found so far is hydro-electric. The Hoover Dam captures solar power by harnessing the Earth's own water cycle, a low pressure, high-volume, solar-heated steam engine. Until the 1930s, most of our electricity was hydroelectric, because until then hydro remained cheaper than thermal power from fossil fuel. But dams disrupt a lot of land and water, both upstream and down. Soft Greens hate them.

It was by learning to take more of what lies far beneath the land that we learned—only very recently—to use so much less of the land itself. This is the most important ecological truth in all our environmental debates about energy, and it is the one Amory Lovins got horrendously wrong. The best way to save the biological environment is to do things *away* from the surface, away from places where the living things dwell. The best way is to do the dirty work in the third dimension, rather than doing it in the first two.

In the last century or so, that is just what we have begun to do. Digging deep has done much to save what remains of our hardwood forests. Plastics, cement, and steel have replaced hardwoods as our main construction materials. Environmentally speaking, the high-rise apartment building in the city is doubly frugal: it houses many people on a small area of surface, and it is built largely from materials extracted from the dead depths of the earth, rather than materials harvested from the living surface.

Mining the depths has delivered even greater environmental benefits in connection with food and energy. We use less land to grow food because we convert oil into solar-power-enhancing additives: fertilizers, pesticides, and plastic packages. And still less land because we bioengineer high-yield crops, develop growth hormones for our cows, use better preservatives, and irradiate our food. Preservatives, packaging, and irradiation all save food, and thus save land. So do frost-free strawberries and genetically engineered tomatoes.

End to end, we keep increasing the efficiency of our agriculture, which means we extract more solar energy out of less acreage. Technology historian Jesse Ausubel calculates that worldwide, a land area the size of the Amazon basin has been spared since 1960 simply through the improvement of agricultural yields. Aquaculture, a booming new industry, is the energy-intensive, feed-rich cultivation of fish in ponds, artificial lakes, and ocean cages. Americans are eating more fish than ever before, but we are getting it out of far less water, with far less harm overall to oceans. By 1800, whale populations were in precipitous decline. Greenpeace didn't save the whales, Colonel Edwin Drake did. Leaving the Arctic Ocean to others, he mounted his harpoon on a derrick in Titusville, Pennsylvania. He began pumping oil commercially in 1859.*

---

*That same year, an ocean away, Charles Darwin published *The Origin of Species*.

So far as the rest of our energy supply is concerned, we are feeding off the fat of the coal, the oil, and the uranium. What we mine from the depths of the Earth now substitutes directly for what we would otherwise have to reap, harvest, gather, scrape, and flood from a vast area on the surface. Coal and oil deposits contain a lot of energy because they represent thousands of years of captured solar energy. Coal is dead trees, fossilized biomass, solar energy refined first by nature, then by geological process. Eight hundred million years ago the Earth's air was mostly carbon dioxide. Green plants evolved and flourished in such profusion that they sucked up most of it. Some sank into swamps and sank deeper. Coal is simply yesterday's landfill, all the stuff that didn't get recycled or composted. Uranium—radioactivity—is "solar" energy of another kind, even more elemental. It was formed through the same condensation of matter as the sun itself. Tiny volumes, extracted from a truly minuscule amount of real estate, could power all human activity on the planet for ages to come. Lovins is right about one thing: When we use those fuels, we are indeed living off "capital": *solar* capital, dead capital, immeasurably vast amounts of capital condensed or buried deep in the Earth at the creation itself or over billions of years thereafter.

Even mining and extraction affect the local environment, of course, which simply means that we should favor places and means that affect it the least. The more energy rich the fuel extracted the less disruption there will be. Generally speaking, the greenest fuels are the ones that contain the most energy per pound of material that must be mined, trucked, pumped, piped, and burnt. Coal is the least good hard fuel by that standard, oil is significantly better, and nuclear is millions of times better still. Strip mining is more environmentally disruptive than shaft mining, even if shafts are more dangerous for the miners. It is better to drill for oil in a desert, or over permafrost, than in a Louisiana bayou. There is a lot less life in those places to disturb.

We use considerable amounts of all these hard fuels. Extracting comparable amounts of energy from the surface would entail truly monstrous environmental disruption. Light and heat, and their immediate by-products, biomass, wind, and rain, are thin, diffuse, and difficult to harness, by comparison with hard fuels. To capture even a modest amount of such forms of energy requires a great deal of . . . *land*. You have to occupy a lot of land, a lot of surface, and you have to scrape it clear, again and again. Yes, you can live off this current energy "income" all right. It is the income that a wheat farmer captures in Idaho and that a cassava farmer captures in Brazil. But endless miles of wheat are not biodiverse prairie. They are "green" only in the most superficial and misleading possible way.

So what do the Softs advise? Not another Hoover Dam, exactly, but its environmental equivalent. Amory Lovins exhorts us to live off the land once again. We are to grow vast plantations of guayule. We are to become Soft, post-industrial Bushmen, hunting and gathering our energy where we can find it, just as we used to. We are to burn wood, garbage, bacterial mats, sunflower oil, buffalo gourd, peanut shells, chicken droppings, and tallow from lambs. Or if we get ambitious enough, we may build the Hoover Dam again, but we have to build it (at enormous expense) fifty miles off the Scottish coast. The world's first commercial wave-power station was indeed built there. It sank in a storm in 1995 less than a month after it was launched.

To Hard Green eyes, the prescriptions of the Softs are grotesquely perverse. However good the intentions, the results are wantonly destructive of the environment. Stop using premium fuel in big power plants, Amory Lovins instructs. Prefer instead the fuel that led to the deforestation of all England, the fuel of the Third World today: biomass. Cow dung, in India. Wood in much of South America and Africa. And for the west, guayule or chicken excrement. The Hard Green recoils. This is a prescription for destroying land, not for saving it.

Everything Lovins would have us use for energy is land intensive. It has to be, because the only current energy "income" the Earth has is solar energy, and solar energy is very diffuse. Softs say that rooftops provide free acreage, but they rarely do, for a wide variety of reasons. Not the least of which is that in the city, where the most people dwell on the least amount of land, there is far too little rooftop to do the job. And because they are land intensive, Soft energy sources are the very antithesis of green. Farming wind is even less green than farming wheat. Wind farms litter the landscape, they kill birds, they are horribly noisy, and they generate only trivial amounts of power. Nobody talks about growing guayule any more: It requires vast amounts of land, yields little oil, and causes debilitating skin rashes. Wave machines generated ear-piercing screeches, but no useful power.

It is just as bad with food. "Organic" food requires far more land to produce—and is therefore far less green—than land-frugal, factory-farmed alternatives. The modern pursuit of absolute "purity" in our food is pleasant, for humans. But it takes more out of the rest of the biosphere, not less. The whole back-to-nature, farmer's market theory of Soft

Green, the entire psychological infrastructure of the movement, is anti-environmental. Taking five billion humans "back to nature" is the worst possible thing we could do, not only for the humans but for nature, too. It is when Monsanto develops a new pest-resistant corn that doubles the farmer's yield per acre that something really is saved: land itself. Same with a growth hormone that delivers more cow and milk on less pasture.

Imagine, now, what would happen if the acolytes of Soft somehow did develop a fantastically hardy and productive guayule plant, one that could grow oil for us in our very own forests or prairies more cheaply than we could drill it out of the desert in Saudi Arabia. That would certainly be the end of our forests and prairies; they would all be plowed and planted in short order. Imagine a super-efficient aquatic weed that might do the same at the edge of the ocean, a super-Sargasso sea, producing oil even more cheaply than a derrick. There would soon be millions upon millions of acres of the weed under cultivation, instead of coral reefs and all else that dwells there now. Imagine what would happen if solar cells ever became "too cheap to meter." We would plaster our rooftops with zero-cost solar collectors, but they wouldn't come close to powering the lights and computers in a high-rise. So we'd cover Kansas and Iowa, too, pretty much as we do now, when we grow wheat and corn, the cheapest collectors of solar energy currently at hand. The more energy we needed, the more surface we would cover. The effects would be about as benign for the environment as strip mining the continent and paving it with asphalt.

Soft Greens would have us live once again off the surface. Hard Greens are bottom feeders. Softs advise us to consume live fuel. Hard Greens dig for dead. Soft energy sources consume the surface of the land. Hard ones consume what lies far beneath it. They are Bushmen, we are troglodytes. With food and energy, the Softs prescribe the environmental equivalent of suburban sprawl; Hards prescribe the equivalent of living in the city. Lovins, an American expatriate in Britain during his intellectually formative years, never grasped the historical lesson of that small island. "Soft" energy sources are horribly land intensive. England was Soft until the 1600s. If it had gone Hard a century or two earlier, it would still have its great forests. But it didn't. The English did instead precisely what Amory Lovins advises us to do. They lived off the surface of their land, not its depths. They scraped and chopped their fuel from the surface, wherever they could find it.

Happily, we Americans changed course before we had completely deforested our own continent. And happily a second time, we gave Lovins a polite hearing and then forgot all about him. We stuck with the Hard path. We stopped gorging on the fat of the land, as Lovins urged us to do. We continued to do our gorging, instead, far below the surface.

Doing things in the third dimension is difficult, but it is also green. Doing them in the first two, though said to be Soft, is more difficult still, and less green. The best we can do to conserve life at the surface is to find our energy, and do our traveling, somewhere else—in mines, tunnels, or jets—anywhere but on the surface where the life is. It is far more environmentally friendly to live in the three-dimensional city than in the sprawling two-dimensional suburb, in the high-rise rather than on the ranch. Planes use a lot of energy, but they are kind to the land, far more so than trains, which devour real estate. The greenest possible strategy is to mine and to bury, to fly and to tunnel, to search high and low, where the life mostly isn't, and so to leave the edge, the space in the middle, living and green.

SCARCITY

But we will run out of third-dimension stuff, the Softs reply, because it is not self-renewing. Oil, for example. Of course we will run out, some day. Malthusian doom can be predicted in three dimensions as easily as in two. But all experience teaches that before the doom arrives, human ingenuity comes up with something different. Uranium alone would serve as a complete substitute for such a long time to come that when that distant day arrives we might easily be gathering energy from the vicinity of Alpha Centauri. As I discussed in Chapter 1, *The Limits* modelers simply fail to confront the sheer vastness of the Earth itself. Weighed in three dimensions as it must be, the planet is truly huge. Its resources remain unmeasurable but vast.

On another front, Soft Greens now foresee a looming scarcity of genes. Species are being extinguished by man's relentless expansion across the face of the planet. While one may argue about the details, the trend surely is in that direction in most countries, although not in the United States. But even here, the Soft Greens muddle the diagnosis and obstruct the cure. We will need the genetic stock of the rain forest in future years, the Soft insists, to harvest cures for cancer or disease-resistant food crops. But genes are just complex chemicals. Darwin's Invisible Hand concocted the particular mix of DNA found on the surface of the planet at this moment in evolutionary history, not Adam Smith's. Except, that is, in kennels, stables, and experimental farms, in fermentation tanks and gene labs, where capitalists readily cook up genetic diversity for themselves: Chihuahuas and St. Bernards, frost-free strawberries and disease-resistant corn, and transgenic pigs. And it is surely greener, all in all, to cook up genes in a Genentech laboratory than to harvest them from the upper reaches of the Amazon. *That* path to genetic diversity, much reviled by the Softs, is the only one that lets us leave the rain forest alone.

It is the rise of the sandpile, the rise of technological complexity, that averts scarcity of everything: food, energy, genes, life itself. The only limits to how much food we can grow, energy we can extract, houses we can build, miles we can travel, pigs we can breed, diseases we can cure, are the limits of human ingenuity. And they keep receding. The Malthusian limits will never arrive.

The Faustian limits might. Globally, the one real, growing scarcity is scarcity of green. Theodore Roosevelt saw that a century ago, and we certainly see it today, notwithstanding the real progress we have made. We are all the more conscious of green scarcities today because our love of the wilderness has grown along with our material wealth. As our ordinary needs are satiated, we crave all the more the things that T. R. cherished a century ago, the hidden spirit of the wilderness, the silent places, unworn of man.

The Softs see scarcity everywhere, scarcity of wood pulp, corn, aluminum, and oil. So they hector us from every side to husband and squirrel, to conserve and recycle. The Hards discern only one scarcity, scarcity of wilderness, of untouched forest, lake, river, shore, and ocean. The Hard Green is not distracted by a thousand phantom scarcities of dead things that are easily extracted from deep in the Earth. He focuses on the one great scarcity that matters, the looming scarcity of wilderness and wildlife at the interface.

EXTERNALITY

Many conservation objectives today involve things more fluid than forests, clean rivers and clear air being paramount among them. Such objectives are as much a part of the Hard Green agenda as of the Soft; we part company only when the pursuit shifts from visible green aesthetics to trans-scientific phantoms.

Saving the Earth—the Hard way—is the cheapest, most effective, and most pleasant way to control pollution. Conservation—the T. R. kind—

is by far the best way to purify air, water, and earth. As I noted in the last chapter, the most beautiful way to purify water is probably the most effective too: maintain unspoiled watersheds. The most effective way to suck carbon out of the air is to grow trees.

The most cogent Soft objection to a fossil-fuel economy is that if we dig up and burn all the fossilized plants of the Carboniferous period, we can expect to recreate the atmosphere of that period too: a carbon-rich hothouse. So what, the hardest of the Hards reply? Carbon dioxide is a nutrient for plants, so the planet will just get greener still. The rain forest will come to Idaho. And as for Miami being under twelve feet of water, well, it is curious to hear Softs opposing what's good for the rain forest because of what's good for the metropolis. But few Hards are *that* hard. Writing off Miami is a serious business, and transforming the global climate more serious still.

As it happens, however, it won't happen, at least not if we follow the Hard road where it naturally leads. As humans leave the land they once cultivated, trees return. And if we resist the "soft" urge to cut them down for fuel or to grow corn for methanol in their place, the trees will simply . . . grow. And remove carbon from the air as they do. The mature rain forest is of little use for that purpose: The greenery sucks in carbon all right, but rot, mold, and bugs on the forest floor work every bit as hard as the trees, composting dead leaves, biodegrading everything in sight, and sending carbon dioxide straight back into the air. But *new* growth *does* sequester carbon, and lots of it. New trees are where the carbon in the coal began. New trees are the first place to which a lot of it can return.

The numbers already bear this out. America releases 1.6 Pg of carbon per year into North American air by burning fossil fuels. (A Pg, or "petagram," is a million billion grams.) Prevailing winds blow west to east. This means carbon dioxide concentrations should be 0.3 parts per million (ppm) higher in the North Atlantic than in the North Pacific. But in fact, they're about 0.3 ppm lower. That last number isn't theory, it's direct measurement. All in all, North America doesn't dump carbon dioxide into the air. It sucks the gas out. Bottom line: America's "terrestrial uptake" of carbon runs about 1.7 Pg per year, just ahead of our fossil fuel emissions. The rest of the world doesn't keep its carbon books in balance. But America does.

How do we do it? An October 1998 article in *Science* summarized the Hard answers.* First, there's "regrowth on abandoned farmland and previously logged forests." Growing new trees sucks up a lot of carbon. Hard agriculture produces so much food on so little land that we are returning lots of old farmland to forest. Second, there's uptake "enhanced by anthropogenic nitrogen deposition." That's nitrogen "pollutants" and fertilizers, for the rest of us. Third, we owe some of our carbon uptake to the coal miners themselves, the culprits in chief. What *Science* obliquely refers to as "CO_2 fertilization" promotes green growth, just as phosphates do. Up to a point, greenhouse gas is its own antidote; nature sees to that. America, in short, has reached a unique stage in its development, high technology, and land management. Alone in the world, we're reforesting and fertilizing enough to suck more carbon than we blow. We recycle our carbon. If greenhouse gas is a problem at all, the rest of the world is the problem. America is the solution.

Carbon illustrates how easy it is to overlook the green upside of Hard technology. Few Softs spend much time defending the chemical factories that manufacture ammonia or the mining companies that extract phosphates. Yet mining phosphates for fertilizer, it turns out, is a direct antidote to mining coal for fuel. There are better ones, of course. Coal itself is a centuries-old fuel. Newer, harder technologies are cleaner still.

Much of the time, as I discussed in Chapter 2, it is our own past, the legacy of yesterday's market and of nature itself, that makes the problem of pollution so difficult. Externalities are everywhere, they surround us. We can't abolish them instantly, and we can't abolish them completely. The best we can do is migrate toward cleaner alternatives, and that means tolerating some measure of new externality pollution to displace a larger measure of old. As Harvard physicist Richard Wilson has demon-

---

*

Carbon Budget	Fossil emissions (Petagrams/year)	Terrestrial uptake (estimated range)
North America	1.6	1.6 to 1.7
Eurasia and North Africa	3.6	−0.4 to 0.5
Tropics & Southern Hemisphere	0.7	−1.1 to 0.9
Total	5.9	0.1 to 3.1

S. Fan et al., "A Large Terrestrial Carbon Sink in North America Implied by Atmospheric and Oceanic Carbon Dioxide Data and Models," *Science* 282 (October 16, 1998): 442–446.

strated, blocking almost any new power plant is anti-green, simply because nine-tenths of every "new" plant is not new, it is just a Taurus plant replacing a Model T, cleaner, more efficient, better all around. An older facility gets retired the day the new one opens.

The opportunities for such trade-offs are multiplying, and the gains are accelerating. The microchip and the communication revolution are changing everything. The advance from T to Taurus took seven decades; in many industries, cars among them, we now expect comparable advances in seven years. Such advance means more comfort, convenience, and efficiency, more wealth . . . and more green, too. A central plank of the Hard Green manifesto, then, must be to facilitate that dynamic. As Schumpeter observed, capitalist progress is itself a process of creative destruction, of tearing down what was to make room for what will be. Green progress must follow a similar trajectory. The greenest policy is to step out of the way of that process, not to obstruct it.

This doesn't mean pouring public money into one or two technological darlings like solar cells or electric cars; it means letting private money find its own path to progress. The way to liberate technology is to liberate capital. Capital is what fuels technological advance. High technology is capital piled upon more capital, it is the confluence of wealth and intellect. Capital is what lets us tear down old markets and replace them with new ones.

EFFICIENCY

Still, energy conservation is surely better than waste—the Soft Green can at least remain confident about *that*. However scarce or otherwise our energy resources, whichever ones we use, it is still surely better for the land if we use what we use efficiently. A penny's worth of energy saved by a more efficient refrigerator, lawn mower, or car, or of paper or wood or aluminum saved by recycling, is surely an inch of land saved back at the oil well or mine.

The Hard Green's mildest response is that the amount of surface disturbed by wells and mines is tiny compared with the amount of surface they allow us to save with fertilizers (less farmland) or planes (less highway and rail track). And because their principal dimension is down, into the Earth, the amount of surface these systems consume does not rise at all steeply with the amount of material or energy they extract.

But the Hard Green can respond a lot more strongly than that, and he must. Even efficiency—efficiency of the kind the Soft pursues and prescribes through myriad, meddlesome government programs—is bad for the environment, not good. Finding new ways to live in the third dimension is sometimes efficient, sometimes not, but it is always better for the land on the surface. The Soft Green preoccupation with gas guzzlers, refrigerator compressors, and "energy conservation" in general is distracting, at best. Much of the time it is positively harmful.

As I have argued already, efficiency is surely not green in and of itself. Strip mining is more efficient than shaft mining; more net fuel is extracted by stripping from the surface than by tunneling far beneath it. But for all its efficiency, strip mining destroys more land. In a strict, thermodynamic sense, nuclear power is highly inefficient: A reactor extracts only a tiny fraction of the energy locked up in the uranium atom nuclei. But nuclear nevertheless extracts enormous amounts of power from minuscule quantities of land. When it does reduce mining, lumbering, and such, as it occasionally (though not often) does, recycling has a positive environmental impact, even if the recycling trucks run on oil.

However easy it is to confuse the two—and Soft Greens confuse them all the time—efficiency is not the same as frugality. It is easy to suppose that every time we buy a little extra efficiency, in a refrigerator or a car engine, we are saving some coal back at the mine-head, just as it is easy to imagine that we reduce our caloric intake every time we pop the tab on a Diet Coke. This is the trickle-up theory of green, the vague logic that what you save in the more efficient refrigerator must translate into savings at the power plant, which must translate into savings at the mine-head.

But as I discussed in Chapter 4, the trickling-up is by no means inevitable; there are too many other directions in which the savings may trickle. The money you save from the more efficient compressor in your refrigerator can easily trickle into a bigger refrigerator, or into a winter vacation in Aspen. It generally does, just as the Diet Coke generally trickles into a brownie. The upshot is no savings at all. We insulate our houses better, build more efficient furnaces, refrigerators, washers, and cars, and still end up consuming more electricity, more carbon fuel, not less. Efficiency is green only when it definitely and directly lightens our tread on the land and its environs. It rarely does. The new *Titanic* is far more efficient than the old, but building and filling another cruise ship, however efficient, doesn't save any earth or ocean.

The only certain effect of efficiency is to increase wealth—and it does that only when the efficiency comes about through market forces and personal choice, not coercion. In the production of goods and services, efficiency enriches the producer. This is why we hardly need government edicts to promote efficiency among producers: They are quite eager enough already to promote it on their own. The builders of the new *Titanic* didn't have to be ordered by the government to replace the old design's inefficient coal furnaces with efficient new diesel engines, nor to add a porpoise's nose to the hull, at the front, below the waterline. They thought of it all by themselves.

The same holds true among consumers, the only difference being that consumers maximize personal satisfaction rather than profit. And satisfaction is, of course, a lot more personal and quirky than business profit. Sometimes it is downright perverse, environmentally speaking. Some people are pleased to do green things like biking or walking; others take pleasure in driving too fast in a 1968 Mustang convertible. It is meaningless to speak of "efficiency" in this context; for those who take direct pleasure in profligacy, as for those who just prefer not to bother about saving every last possible dollar in their energy budgets, "inefficiency"— the thermodynamic kind—is often "efficient," in economic terms. The only overall effect of forcing such consumers to buy "efficiency" or "conservation" that they wouldn't buy otherwise is to make them a bit poorer.

Setting aside land for "conservation," as T. R. did: *that* saves the Earth. So does promoting technology that moves human life into the third dimension, the dead one, and away from the surface, where the rest of creation dwells. But "efficiency," "energy conservation," and such directly affect wealth, not the environment. Efficiency—voluntary, by choice, impelled by market forces and free consumer choice—has only one definite, predictable effect: It boosts wealth. Coerced efficiency—efficiency impelled by dictate, edict, fiat, regulation, and central control—has only one predictable effect, too: It lowers wealth. Coerced efficiency, like coerced recycling, is simply another instance of doing things well that should not be done at all.

DEATH IN THE THIRD DIMENSION

However efficiently we use it, extracting all that fuel from the bowels of the Earth creates wastes. Wastes, too, are an important part of living in three dimensions. Most of the time, the best thing to do with our copious wastes is to bury them. With rare exceptions, recycling is the worst possible option.

To illustrate, consider once again the carbon cycle, at its tail end. The carbon sinks we get by reforesting millions of acres of old farmland are secure. But the sinks created by all the ammonia-fertilized, phosphate-fertilized corn and wheat don't remove carbon from the air for long if the crops end up in a grasshopper's gut. Guts and grasshoppers reverse carbon flows, pumping carbon dioxide back into the air. So pesticides are gas-busters. So are termite exterminators.

Our own voracious appetites can take care of the rest of the excess carbon, and already do, except where misguided Soft policy gets in the way. Our own growing bodies sequester a bit of carbon, of course, though most of that returns eventually to the air unless our coffins are sealed very tightly. But the one thing a fat society does persistently and well is generate copious amounts of carbon-rich waste. About two-thirds of what we put in landfills is carbon based. And buried in a modern landfill, the carbon goes . . . nowhere. It doesn't rot. The landfills are well compacted, their contents stay dry. They are not composters, they are mummifiers. University of Nebraska Professor Craig S. Marxsen calculates that "appropriately constructed landfills could capture roughly 2 billion tons of carbon annually, right now, and virtually stop global warming cold in its tracks."

By mummifying carbon, we simply complete the carbon cycle. For a society that is consuming 70 quadrillion BTUs of fossil-fuel energy every year, there is only one honest way of "recycling" carbon wastes, and that is to put them back where most of the carbon we use came from, deep underground. Composting food wastes and recyling newspapers are the last thing we should want to do: Both interrupt the return of carbon to the Earth. The notion that "there is no room" down there is absurd. If we take old carbon out of the ground, we can put new carbon back in. Only two-dimensional thinkers could possibly believe otherwise. In three dimensions, there is always plenty of room.

THE SANDPILE IS GREEN

Up to a point, we have simply been lucky to have all that coal and oil and uranium, all those dead-and-buried geophysical resources available as substitutes for living resources on the surface.

But the fact that humanity is now able to live in the third dimension is not just an accident of geophysics or of technological history. Man threatens all the rest of life on the planet because his frontal lobes have made him so adaptable, so very fit to survive, that he is now capable of surviving on the moon itself. Our ability to live off resources long dead and buried—resources that are wholly unusable by all other forms of life on Earth—is merely the logical, inevitable, extension of our ability to subdue the biosphere. We are very clever indeed, clever enough to cut down all the forests of England for our stoves; clever enough to hunt whales into extinction for lamps; and now clever enough to stop destroying those ancient, living, surface-dwelling sources of fuel by aiming our steel harpoons and diamond drills at dead things rather than at living ones. The technology that empowers us to destroy all life on Earth is the same technology that empowers us not to.

Complexity—the sandpile—is green because it lets us do so much more with so much less and to do so much more of it away from the surface, in the third dimension. The ineffably complex technology of atom and photon is the greenest of all. Nuclear power, the original "solar," extracts limitless energy from the tiniest amounts of material because it extracts subatomically. Irradiation is an excellent preservative because its energy disrupts the chemical bonds we most want disrupted, the ones that the salmonella bacterium most needs to live. Improved genes in rice or a tobacco leaf substitute for a chemical factory producing phosphates or fertilizers or a pharmaceutical plant manufacturing drugs. Pesticides that disrupt the hormonal systems of insects, like the hormones that promote growth and milk production in cows, are extremely efficient and extremely safe, too, because they are so accurately aimed at such specific molecular targets. In medicine, the best drugs are likewise the ones most meticulously engineered to affect specific cells or specific proteins on the surface of the cells. Nobody dismisses complexity as "unreliable" or "brittle" in *that* context. We know it is just the opposite.

As I discussed at the outset of this chapter, most of the main objections to Hard fuels center on sandpile scenarios, or more simply on pollution. The coal's effluents will surely spread, the oil tanker may founder, the reactor melt down. Yes, they may. But if we are careful, they won't, and experience teaches that they generally don't. The Soft Green alternatives, by contrast—the guayule growers and chicken farmers and solar freaks— spread out and choke the land by definition and design. They have to, because they are gathering a very diffuse form of energy, and to gather any significant amount of energy that way you simply have to occupy a lot of surface. The Soft technologies cannot collapse like sandpiles because they are collapsed to begin with, and with dreadful environmental implications. The Hard, Faustian technologies are immeasurably better for the land because they are concentrated, because they confine a great amount of energy in a very small space. Their best green qualities, in short, inhere in what the Softs most passionately oppose.

From the perspective of the biosphere, nuclear power is the "simplest" of all in that it disturbs the least land, interferes least with the surface, and corrupts life the least, not the most. It is far more brittle, unstable, and dangerous, all in all, to cultivate endless miles of corn for gasohol, or guayule for oil, or to overlay the prairies with solar cells, because so much of the underlying, green complexity is disrupted by these Soft alternatives. Hard technologies are generally stable, safe, and green because they use more capital and concrete and less land, less biosphere, less nature itself, the highest complexity of all.

The greenest possible policy is to do the exact opposite of what the Softs advise. They advise us to burn live trees, not dead ones. To scrape the surface, not plumb the deep. They could not possibly have been more wrong. Amory Lovins's own little (adopted) island, beautiful England, the land of April and hedgerows and birds, lovely though it still is, should have taught him as much. It was once lovelier still. It once had forests. Our land still does. We should not make the English mistake. We should dig up our energy, bury our wastes, fly high, tunnel deep, and leave more of the surface alone.

8

Privatizing Pollution

———— ◆ ————

Save surface, save wilderness, but pollution remains a refractory, pernicious problem. Private actors smoke, seep, and dump their wastes into air, river, swamp. Public ones pursue ephemera down into the recesses of the microcosm, where pursuit is more wasteful than waste itself. One side impoverishes us by fobbing off serious pollutants; the other impoverishes us by its interminable crusades against phantasms. The private sector's externalities erode public goods; the public sector's erode private ones. One side "pollutes," the other "takes." Both sides muster vast legions of lawyers and consultants to defend their impositions. The upshot, far too often, is a wealth-devouring stalemate, a steady degradation of both environment and economy.

The pollution problem is usefully divided in two: means and ends. It is best to begin with means. Because with pollution, there are in fact no ends.

PRESCRIPTION

Suppose we agreed, both Hard and Soft, that sulfur or nitrogen oxide emissions into our common air are too high for our health, or the pine forest's, or just too high for the air to stay as clear and sweet smelling as we like it. In the early days of the Clean Air Act, we did roughly agree about that. A vast compendium of regulation soon followed. The EPA prescribed scrubbers for coal plants and catalytic converters for tailpipes. Sometimes, when it felt indulgent, it merely directed particular sectors of particular industries to reduce emissions to particular levels, leaving to them the choice of technology.

Such approaches were horribly inefficient. And while they remain on the regulatory books, they have few defenders left today in any green camp. There is no good reason to require every individual polluter to do exactly the same thing; what matters is emission reduction in the aggregate. With pollution control as with everything else in the market, efficiency lies in an endless series of trade-offs, between one provider and another, old technology and new, today's production and tomorrow's. Prescriptive regulation allows no such flexibility. It collectivizes the control of pollution in much the same way as central planners used to collectivize the baking of bread. The end of such planning is always the same: More and more is spent achieving less and less.

SUBSIDIES

Collectivizing the pollution control budget doesn't help, either. Consider the problem of household trash, today's iconic mother of all externalities. Public collection and disposal services were originally established as a public solution to an obvious environmental problem, human garbage piling up where it shouldn't. So today, from the householder's perspective, the dump is free, and the trash collection is, too; he pays property taxes, not trash fees. But as externality economists have been telling us all along, there's no such thing as a free dump. It only looks that way when the costs are buried in the wrong place—in property taxes—and thus disconnected from how much trash each household generates.

What ought we to do? Cut the property tax, eliminate the subsidy, and replace it with . . . nothing. Nothing under government auspices, anyway. Because it is solid and visible, household trash is not a difficult problem for markets. Householders won't just leave their garbage in the streets; the laws against that are strict and easily enforced. Commercial establishments already take direct financial responsibility for the removal and proper disposal of their own trash, and those markets work fine. The trash crisis, such as it is, just replays Hardin's "Tragedy of the Commons." Government, the creator of the commons, is the problem. Less government is the solution.

Many municipal governments have resolved to try more government instead. Rather than get out of trash, they have gone into real estate, glass, scrap metal, and pulp. They call their expanded trash business "recycling."

It is said to save land and raw materials. But perfectly good markets already set prices for those resources. And unfortunately for the recyclers, these things cost a lot less than Softs suppose. New York spends $200 more to collect a ton of recyclables than it would cost just to bury them properly. The city pays another $40 to get rid of the "resources" culled by this recycling effort. The price of recycled newsprint rarely rises above zero; private recyclers demand an additional fee to remove the ink, which costs more than the recycled pulp is worth. Only the most shameless fiddling of the books allows a few other cities to claim that curbside recycling programs are cheaper than burying. If land and transportation were more expensive, if environmentally secure landfills were costlier to maintain, the numbers would be different. But they aren't. That simply tells us that trash is not an environmental problem, not in this day and age, the odyssey of the Long Island trash barge notwithstanding. The barge had nothing to do with externality and everything to do with junk television.

If there were private profit to be found in modern trash, it would be found without government compulsion. Recycling used to be a huge industry and remains so in poor countries. In *Our Mutual Friend* Charles Dickens tells the tale of a man who made his vast fortune in what was then called "dust." A vibrant rag trade flourished in the United States until well into this century. Entrepreneurs, not regulators, established the first large wastepaper mill in the United States, in 1913. In many countries to this day the very poor set up their shacks alongside dumps, living off what they can extract from them. In times of desperate famine, starving African children pick through human excrement in search of undigested kernels of corn.

But such things do not happen in affluent countries, and for good reason. These stories are not about the tragedy of the commons, they are about the tragedy of life. They are about human poverty, not environmental decay. In America, we do not have enough poverty for such things to happen spontaneously any more. So, perversely, the Softs try to substitute politics and propaganda instead. The upshot isn't more green, it's less. We would make things greener faster if we simply took the money and effort wasted on recycling and used it instead to expand publicly owned parks, forest, lakes, rivers, and shores.

Less government is, in fact, the solution to lots of the scarcity problems that so concern the Softs. Municipal government supplies free water, so we waste it. Government subsidizes highways and public transportation, too, nukes and windmills, oil and ethanol, hydroelectric dams and fluorescent bulbs. Softs may agree to cutting subsidies on a few technologies like nuclear power that they particularly revile, but invariably so as to boost others as much and more. When markets fail, Softs prescribe government, and when government fails, they prescribe more government.

Of course they do. They hail from the political Left, and the Left's instincts lean toward expanding the public sphere, not shrinking it. Softs of the Left believe in the commons, believe it does more good than harm. The Left declines to avert Hardin's Tragedy by shrinking the commons itself.

TAXES

Anathema to the Right, however, is the prospect of the commons taking over everything, not just the public park but all of private life and property, too. The Right remembers that the Communists had nothing but common property and laid waste to it all.

That is why the politically savvy Hard Green won't touch the "Great Green Tax Shift," promoted by otherwise sensible economists like Paul Krugman. The idea is simple enough. Many of our current taxes are imposed on good things, like employment and labor. We should impose them instead on bad things, like pollution. Soft Greens used to denounce such taxes as immoral, but they have gotten over that, and today they would be delighted to enact one. They fail, says Krugman, because Congress panders: to populist demand for cheap gasoline, to labor pressure to protect coal-mining jobs, and to "conservatives who, unlike most economists, really do think that the free market is always right." The Right, Krugman argues, just won't accept new pollution taxes, not "even with the assurance that other taxes will be lowered at the same time."

But Krugman is quite mistaken, about the Right at least. No serious conservative believes that "the free market is always right." Many just recall that Ponzi gave Krugmanesque assurances, too. Washington taxes air travel, assuring us it will build airports, and payrolls, assuring us it will build retirement funds; the assurances are right there, smack in the law, but the airports and the funds are not. And as we have seen, there are plenty of places where *cutting* taxes would do direct environmental good, too, by forcing the elimination of subsidies for agriculture, energy, transportation, and trash disposal. "Show me first your penny," replies the conservative pieman to simple Krugman. With no penny forthcoming, the conservative knows that new taxes mean new Commons, which means new Tragedy in the offing.

POLLUTION AS PROPERTY

Such differences can be bridged, some of the time. It is possible to address most serious pollutants in ways that do not expand the public sphere and that don't undermine established private rights, either. It is possible, in other words, to address the problem of pollution in ways that don't move the boundaries between private and public spheres by a single inch, in ways that neither expand the private sector's right to pollute nor expand the public sector's power to regulate. The upshot is less private pollution and a diminished public sector, too.

Here, to illustrate, is how it might be done with cars and our common air. Your old Chevy pours noxious fumes out the tailpipe? So be it, we shall measure exactly how bad the fumes are and give you a formal permit, and you may continue pouring for as long as you hold it, and you may hold it forever. You now own your right to emit fumes as securely and officially as you own the Chevy itself. Same for all the other cars currently on the road. That is all there is to it. Cars will quickly cut back their grazing on the common air and keep on cutting; the air will get cleaner and cleaner.

How does officially permitting bad things get rid of them? It doesn't, until you go shopping for new wheels. The status quo has been affirmed for the common air, too. The status quo does not include next year's Chevy. That car does not have a permit for its tailpipe, and it needs one to get on the road. So it will have to buy one from someone who does. Permits may be bought and sold, and they will be. They are transferable and subdividable, of course.

If you don't believe in markets this all seems like a complete wash: One person stops polluting so another can begin. But markets always do far better than that. You don't buy an extra ton of pollution permit for $100 if you can develop technology that cuts the same amount of pollution for $50. The moment the market is fully engaged, it innovates. In this anti-matter market, this market of tradable *bads*, the more freely people can sell things, the *less* of those things get produced. The market mobilizes to boost output just as markets always do, which shakes out here as a steady drop in levels of pollution. The market cannot create more permits. But it can create direct competitive substitutes: pollution abatement technology, procedures, and strategies. And so it does, driving the price of pollution abatement down steadily. On a going forward basis, for new cars, pollution costs are fully internalized.

The price of pollution permits drops steadily, too. If that seems at all worrisome, it shouldn't: The falling price is the hallmark of success. The price of the permits is set by the price of their direct substitutes, things that curtail pollution, like fuel efficient engines, cleaner ignition systems, and catalytic converters. Moreover, Greens and government alike are welcome to buy permits on the market, too, and simply tear them up. As permits get cheaper, so does the tearing. All the while, there is no shadow of a government "takings" in sight. Transforming old regulation into new property concedes the takings problem completely, and solves it. If Soft Greens regret the concession, they should at least welcome the solution.

Prototypes of such markets, though still far from perfected, already exist. A scheme for tradable emission permits was included in the 1990 revision of the Clean Air Act, and electric utilities actively trade sulfur emission permits at the Chicago Board of Trade. The price currently runs around $100 a ton. Emissions have dropped sharply, the price of permits is running way below original estimates, and the cost of abating pollution has dropped by as much as 90 percent. The most polluting plants are emitting far less than authorized under the old regulatory order. Soft Greens complain that the original targets must have been too lax, but only because they cannot quite believe how well markets work. The Softs are bidding in some of the auctions, too, for tearing-up purposes. Futures contracts have already been authorized and will be written once the market matures. There is serious talk of building markets in a wide range of polluting substances, including oxides of nitrogen and particulates.

Many quite elusive externalities, from sewage to whales, can be internalized in this general way. Wild animals, for example. Privatizing transforms them from pests, or at best mountains of meat, to rich economic opportunities for tourism and trophy hunting. Everywhere and without exception, their numbers multiply and their habitats are preserved as soon as secure titles are issued. A century ago, lawyers viewed the wild animal in a cage as a paradigmatic "nuisance": An extensive body of law addressed your responsibilities if you allowed it to escape. Today, free-roaming game is an asset; the hunter pays dearly to capture it, and entire habitats are saved as a result. Put a trophy price on elephants or bighorn sheep, and animals on the brink of extinction are soon multiplying like Frank Perdue's chickens.

As *Economist* editor Frances Cairncross explores in her 1992 book *Costing the Earth*, property lines can be drawn more creatively than skeptics suppose. It is usually a matter of will, not way. Property rights in radio spectrum and offshore oil leases didn't exist either until government bothered to create them, and then the only surprise was how well they worked, how quickly the scarcities were transformed into abundance. Private markets routinely promote "environmental" interests with "easements" and "covenants"; you can hardly buy a house or condominium without becoming party to a thick binder of them. More recently, the federal government has allowed landowners to make tax deductible gifts of conservation easements to charitable nature conservancies.

Ephemeral though pollution invariably is, no serious student of property law has any difficulty imagining pollution folded into that legal rubric. All property is an endless succession of bubbles in space, or cyberspace, with different people claiming an endless variety of interests in them. Some forms of property protect solid, stationary things—like land—and control the movement of big things (like people) under the legal rubric of "trespass." Others protect fluid, mobile, and intangible things—like air, water, scenery, or cumberland sandworts—and control the movement of smoke, dirt, and sewage under the legal rubric of "nuisance." Property is a bottle of champagne, or the name of the label, or the whole concept of effervescent wine, or perhaps just wine in your bloodstream while you drive home in your Buick. Control of each little bit of turf, physical or virtual, can always be made a bit more—or a bit less—personal and private. Pollution, or its obverse, clean air or water, isn't any less tractable than lots of other even less tangible things that people own and markets trade all the time.

It takes a new bureaucracy to manage new property rights, the Softs protest. It does indeed. Markets trade, subdivide, and aggregate, but there has to be something there to begin with. All property rights entail administrative overhead: title registries for real estate and cars, exchanges and boards for securities, exchange commissions to promulgate trading rules, arbitrators and courts to decide disputes. The more fluid and ephemeral the property, the greater the overhead in defining and containing it. Patents, copyrights, trademarks, and trade secrets are tremendously slippery forms of property that are concomitantly expensive to define and administer. But still cheap at the price. For all its expensive overhead, the privatizing of intellectual property has been a tremendous spur to innovation. The privatizing of pollution has been, too. Much more is possible.

Prescriptive regulation is indeed simpler and more convenient, just as prescribing who produces and consumes bread avoids the tangle and snarl of the free market. Simpler for the bureaucrat who insists on tracking and monitoring every loaf, everywhere. More convenient for the agency that retains complete control over every last grain of production, innovation, and consumption. Prescription is said to be fairer, too, better at avoiding the problem of rich and poor, and its pollution equivalents; some neighborhoods are more fouled than others. Rationing bread is fairer, too; at its best, it spreads the poverty around quite equitably. Central planning is always fairer, at least when the planners remember not to look after themselves first, a detail they sometimes overlook. But all this simplicity, convenience, and fairness is achieved at the bottom of the barrel. It is possible to do far better.

Getting new pollution markets started is always the most difficult part, but happily the details often don't much matter. As Ronald Coase demonstrated, downstream markets will generally see to it that property rights end up in the most efficient place, however you assign them initially, so long as the rights are readily tradable. Once the sparks from the train are owned and traded, either the railroad ends up paying the farmer for burning his crops (if it "owns" the problem) or the farmers pay the railroad not to (if they do), with the overall balance between railroading and farming set efficiently, depending on the relative value of each.

Only people who do not believe in markets at all imagine that free market types strenuously oppose the intelligent privatizing of pollution. Why should we? Property—of any kind—is the free market's friend. Sooner or later the capitalist squeezes a dollar out of it. Much of the time, his interests already align closely with intelligent environmental protection. He does not want to melt down his nuke; he will lose his capital if he does. He might prefer not to have to think about pollution at all, but hand him a tradable permit, and he will instantly begin inventing ways to turn it into profit. The captain of industry who juggles ten thousand pieces of property to build and run a power plant can add another dozen to his books without blinking an eye. He deals with tradable pollution permits the same ways he deals with tradable turbines and fuels. He wheels and deals, improves, innovates, and laughs all the way to the bank. Which is exactly what we should want him to do.

Until recently, Soft Greens viewed permits as even more reprehensible than pollution taxes, altogether incompatible with any moral perspective on pollution. Softs didn't want polluters to trade indulgences, they wanted polluters to expiate sin. But they are getting over it. Politically, Hard Greens aren't giving people anything they don't already have.

To take what they have away from them we need votes, not only in legislatures but in courts, too, to overcome Takings challenges. But the votes aren't there, and they won't be. So the best thing to do is to issue permits for the status quo, but not for anything that will make things worse. Innovation will then depreciate the value of the permits year by year, and they will be torn up in the end. However oxymoronic it may sound, permitting bads is the most effective way to get rid of them.

FIDEL'S BOOKKEEPER

Markets cannot function, however, without honest bookkeeping. That means keeping track of credits as well as debits. Growing new trees removes carbon from the air. So does mummifying organic and plastic waste in landfills. If we are serious about our green objectives, we will keep honest books. If we can't or won't, no progress will be made, not through markets, and not by conventional regulatory prescription either.

Many Soft Greens viscerally reject the whole notion of pollution credits, most particularly for seemingly extraneous acts of green kindness, like growing new trees. But pollution is not sin, and so long as the books are kept honestly, we should *want* polluters to buy their way back into our green graces in the cheapest way they can find. The cheaper the ways, the more we can get them to buy.

Embezzlement is a related problem, and markets themselves cannot solve it. No honest burgher will buy what he or his neighbor is lawfully permitted to take for free. It is no use fencing in just the north side of the Commons, if it is just a short walk around to the unfenced south. America can control pollution in its own rivers and local air whatever happens in Brazil. We should save our cougars whether or not Brazil saves its jaguars. But there is only one global climate, and there is, increasingly, only one global market for steel, petrochemicals, aluminum, and many other energy-intensive, air-devouring, water-fouling industries. There will be no market for greenhouse-gas permits in Alabama if it is just as easy to burn unpermitted fuel in Brazil. Uneven environmental regulation has already moved electric power plants from New York to Quebec and from California to Utah, with high-voltage transmission lines slashing across the country to ship power back to the communities too green to generate their own. This has not done the environment any good at all. Relocating steel factories to China and petrochemicals to Dubai will—quite obviously—harm the environment, not help it.

By the same token, overfishing of the oceans—a serious problem—will not be solved by substituting Japanese dragnets for American ones. Hard Greens would be quite content to see intelligent property rights affirmed in this arena, too, but that takes responsible enforcement authority on the scene, and there isn't any. Hards part company with the Softs when it comes to believing that congenital thieves, liars, and bandits—the likes of Fidel Castro and countless other nasty little despots around the globe—can be relied upon to affirm property, regulatory norms, or anything at all remotely related to the rule of law. We would trust gunboat diplomacy instead, which is just what America trusted when it unilaterally extended its "territorial waters" out 200 miles. Extending them another 1,000 or so would do wonders for the ocean environment. When Soft Greens finally grasp geopolitical realities, this urgent problem will be solved, but not a moment sooner.

All in all, it is a gross intellectual fraud to set about solving a "commons" problem piecemeal, here but not there, north but not south, west but not east, one industry but not another, debits but not credits. The commons problem arises *because* a resource is shared, so that no one acting alone has incentive or ability to take care of it properly. Solutions that aren't equally uniform, equally "common," are a complete waste of time. Half a fence solves nothing, it just forces law-abiding people to take a hike or sell their herds to their lawless neighbors.

GAIA AND THE MARKET

Property rights are the efficient means when man himself must take charge of what he emits. But much of the time he shouldn't.* A scarlet "P" can be branded onto anything, but meddlesome intervention should require more than that. Simple dispersion is quite often the only approach that makes any sense, economically and environmentally, too. To put the matter bluntly, the most efficient way to control many things so tagged is not to worry about them at all. Let them be. Leave them to Gaia.

---

*I'm sure this paragraph will be the one most quoted by the people who least like Hard Green and all it stands for. It would seem to expose us Hards as heedless polluters. Yet no one who is interested in pollution or its effective control can doubt that often, in environmental matters the only rational course is to leave bad enough alone. Things are never so bad that they cannot get even worse.

Softs never acknowledge the detoxifying effects of dilution. They see harms accumulating and compounding regardless, and they put no faith at all in nature's own power to absorb, metabolize, and sanitize. No matter how dispersed pollution may be, the Softs' big computer models undisperse it. They track out every last particle of waste in space and time, add up all the supposed ill effects, reckon up the total cost, and present a bill for payment here and now. The bill is invariably high: It has to be, to justify all the effort that went into computing it. But sufficiently dispersed in space and time, many pollutants do, in fact, harmlessly disappear. Letting nature and planet attend to them is quite often the cheapest, most practical approach, however outrageously that thought may offend the greenocracy. Much of the time we can indeed count on nature's power to dissipate and cleanse. Count on it, because the planet is vast, and because nature is a lot more robust than the Softs are forever telling us.

Noise, heat, and radiation, for example, are all tagged as pollutants, and sometimes they do present real problems worth worrying about. Most of the time they don't. Most of the time, all three forms of energy dissipate quickly and harmlessly. The Earth receives, creates, and radiates such vast quantities of all three forms of energy from other sources that most of the time it would be ludicrous even to think of trying to meter or contain them. Human lungs likewise emit carbon dioxide, and our cattle emit copious amounts of methane, but in those quantities, at least, green plants quite readily take care of our greenhouse gases. Bacteria in fact thrive on our sewage; the main reason not to dump it in rivers is that it breeds so much slimy new life. We just prefer the comparatively lifeless river to the biologically fecund swamp.

As a general matter, neither regulators nor markets are likely to do much good pursuing fluxes of human origin that are completely dwarfed by nature's. Every animal emits carbon dioxide, and every green plant absorbs it. Gaia herself is—by a wide margin—the main emitter of this particular "pollutant," and also its main abater, cycling 100 billion metric tons of it through the atmosphere every year.[1] As I noted in Chapter 7, North America in fact *absorbs* about as much carbon dioxide as it emits, because lumber, agriculture, and natural reforestation take more carbon out of the air than burning fossil fuels emit into it. The prevailing winds move west to east, and average CO_2 levels drop as they do.

Nature is likewise the prime environmental mover of particulates, radiation, hormones, and insecticides. These things mainly originate in, or move under the impetus of, lightning, forest fires, volcanoes, the Earth's core, outer space, and natural life itself. Man's comparatively small contribution to the fluxes and flows may still matter, but neither old-style regulators nor property-centered markets are going to control things when they have dominion over such a small part of the problem. No umbrella will keep you dry when you are standing neck deep in the Mississippi.

BEING AND NOTHINGNESS

Some supposed pollutants, like carbon dioxide, are measured in the billions of tons, others in micrograms. The former are usually simple molecules, and we worry about their chemistry (acid rain) or physics (global warming). The latter are usually complex molecules, and it is their biochemistry that worries us, their power to disrupt hormones or subvert reproduction. Nature manufactures far more biochemically active compounds than man does, but man manufactures great quantities of a few. Some industrial micro-emissions degrade quickly in the environment, others don't. A few, like heavy metals or some long-lasting organic (carbon-based) molecules, can accumulate as they move up the food chain. Molecular complexity permits boundless variation, and there is no telling when the reordering of an atom or two in a large molecule will greatly change biochemical effects. No Hard Green will deny that now and again we stumble across a micro-pollutant that does real harm.

So at what point does a sensible Hard Green become concerned about the levels of a particular toxin in rivers? Are there times when small, trace externalities may turn into "big, obvious" ones in a time too quick for us to react? How do we allay these concerns?

To begin with, some toxic effects are already well enough understood that it makes perfect sense to address them at their sources, where emissions occur, in large quantities and at high concentrations. Arsenic and heavy metals are found everywhere in nature, but mostly buried in the Earth; dumping them at high concentrations into air or water can plainly injure man and the environment alike. Property rights to curtail such emissions are clearly in order. We know pesticides, fungicides, and all manner of other bio-cides affect the environment because we have designed them for that purpose. It makes obvious environmental sense to monitor and limit their use, to contain their disposal, as well as the chemical by-products created in their manufacture.

As a rough rule of thumb, however, it makes sense *not* to start with any presumption of guilt in the micro-environmental realm. Not because micro-toxic effects don't occur; we know they do. But because there are too many possible effects to worry about and too many ways to waste money pursing the wrong ones. For every DDT or mercury, the Softs come up with hundreds of other spurious candidates for very costly control. Micro-toxic alarms are sounded far faster than reliable science can respond, or than a finite economy can realistically hope to address. When science eventually catches up with one or another, the pursuit turns out— much more often than not—to have been a complete waste of time.

Property rights don't help here. No system can fix green problems that science itself is unable to define. It is impossible seriously to imagine a system of tradable permits for the trace toxins that so preoccupy the Softs, the ones addressed in most of the far-future and hyper-dispersion models, the statistical ones, the ones discerned so much more clearly in computers than in the world at large. The harms, if they exist at all, are defined only by layer upon layer of detail in the model. The most complex financial instruments in the market could not begin to shadow what the EPA tries to accomplish with trace toxics, through its opaque, fickle regulatory directives. One industry is regulated completely differently from the next. One country makes completely different calls from the next. Markets cannot emulate something that is idiosyncratic and particularized. The market cannot emulate this kind of regulation, because this regulation cannot even emulate itself. What we know about most items on the micro-toxic regulatory agenda, if we know anything at all, is so specialized, so elaborate, so complex that only the Soft Green priests know how to handle it.

The strong reason for caution about micro-environmental regulation is that it can get so uselessly expensive, so very fast. The pursuit itself can rapidly come to consume more energy, material, and time, endanger more lives, generate more pollution, and dissipate more value, than the thing pursued. Billions of dollars of Superfund taxes are spent digging up acres of dirt and carting it uselessly around the country, doing as little good for the environment as a wayward barge loaded with Long Island trash. The steam-cleaning of rocks and beaches after an oil spill quite literally sterilizes them, at enormous cost to the oil company, and to nature as well. Emissions standards for some chimneys and tailpipes now require the air to be cleaner coming out than when it came in the intake at the front end. Radiation emission standards make it safer, so far as radiation exposure is concerned, to sleep at the perimeter wire of a nuclear power plant than to sleep in bed next to your spouse.*

A robust body of mainstream science now confirms that the benefits of micro-environmental regulation are far more modest than was believed two decades ago, when Soft environmentalism embraced this new regulatory mission. In the visible, countable, measurable world, the numbers are clear. People first: life expectancies increase, birth defects and cancer rates hold steady or decline; taken altogether the things that assail people seem to be losing ground, not gaining. On the environmental front, there is no reason to suppose that the effects have been much different. Many environmental declines have occurred, but most are readily attributable not to microscopic assaults but to macroscopic ones, like bulldozers, asphalt, and fires used to clear forest for farmland. The micro-assault theories are a dreadful distraction here. By and large, we should welcome the industrial changes that trade off macroscopic assaults like those for the comparatively innocent micro-pollutants on which we waste so much money and concern. As we have seen, that means, by and large, welcoming the changes that replace Soft technology with Hard.

Absent convincing scientific evidence in hand before the chase begins, the pursuit of micro-toxics will do more harm than good. Not every time, but so often that there should be a strong presumption against beginning at all. Absent a compelling reason to chase after the invisible, we are far better off pouring our money and effort where clear environmental returns can be realized directly. The tens of billions spent on Superfund would have achieved immeasurably more environmental good had they been spent on buying up green spaces, riverbanks, watersheds, and forests. Actively conserving nature, directing resources to visible green rather than invisible, will do far more for the invisible end of things, too, because nature does have such considerable power to cleanse, detoxify, and regenerate. Dredging pesticide-contaminated mud out of the depths of the Hudson may perhaps do some environmental good. Far more good might be done spending the same money to buy up miles of unspoiled riverbank and setting it aside forever. The pesticides in the mud will leach out or break down on their own, sooner or later, and life will return to the river. It happened in Cu Chi, and it can happen in the Hudson. The riverbank itself, meanwhile, may get irreversibly paved with concrete.

---

*Human bones contain trace amounts of naturally occurring radioactive potassium.

For the Hard Green, the last and most principled line of defense against the micro-environmental frolic is the law of Takings, which Soft Greens have good reason to loathe. The core constitutional principle is simple enough: If government "takes" your private property, it must pay you for it. One can certainly argue about how severe the government interference must be to engage the Constitution, just as one can argue about how severe pollution should be to engage the EPA. But at some point it gets big enough, and at that point Takings law just forces regulators and the public behind them to be economically rational. So long as public benefits outweigh the private costs, it will make sense to continue regulating, even if regulation entails compensable takings of property. What Takings law can do, and should, is draw the line at regulation that destroys more value than it creates. If Softs despise that idea, it is because they know that almost all they do in the micro-environmental sphere falls far on the wrong side of that balance sheet.

Too bad for the Softs: This is where they must be stopped dead in their tracks. Witches were once blamed for environmental harms, too, for blighted crops and sheep dying in the fields, but it is not anti-green to oppose the hunting down of the externalities of witchcraft. It is good for the environment, not bad, to keep the focus on real problems rather than imaginary ones. We have to know when to say enough is enough. If we fail to say that to the polluter, we will end up very polluted. If we fail to say that to the regulator, we will end up very polluted, too. There are two roads to tragedy, a Left and a Right. The trick is to take neither.

WOLF TO WOMB

That a line must be drawn does not mean it is easy to draw. Soft Greens start with every advantage in debates about externalities. Softs can always find some larger space, some more fluid resource or pollutant, that moves faster, that is less prone to be confined in any clean boundaries, less easy to define within a title, permit, or license. However diligently Hard Greens may work to internalize externalities, there will always be something that lies beyond the outer edge of the private sphere. There will always be a Commons. Human activity will always affect it, too, and arguably for the worse. On that one, Soft Greens have the Second Law of Thermodynamics on their side. However tight the casket, the bell always tolls for everyone.

But Second Law aside, exactly the same thing can be said by Hard Greens, too, about the common spaces they cherish the most. Common decency, for example, and degradation by rap lyrics, pornography, vagrancy, and prostitution. Many things called nuisances not so long ago had nothing to do with pollution, they concerned things today's Left would call civil liberties.

Labels like "externality" and its rough opposite, "privacy," settle nothing. The rancher whose land abuts Yellowstone sees a federally protected wolf straying from the park to hunt his sheep. The rancher wants the wolf removed, at once, and by force if necessary. And how, philosophically, is he very different from a woman with an unwelcome fetus in her uterus? Both can speak equally indignantly about autonomy and personal freedom. And each will face an outsider who replies: The space is not yours alone; I too have an interest in it; nature must follow its course.

Then there is the matter of banning the conveyance, through public places and particularly through schools, of certain biochemical products that are almost invariably fatal if they get into your blood. Or if not ban, at least label them very clearly, so that people may keep their distance if they choose. But the toxin in question is not TCE or plutonium, it is HIV, conveyed in the imperfectly secure confines of human bodies. Decent people do not even speak of "externalities" here, they speak instead of moderation and common sense and the importance of not surrendering to paranoia. Yet not so long ago contagious disease could get you forcibly confined. Leper colonies were not established for the benefit of the colonists.

Economists and scientists may help structure such debates a bit, but that is the most they can do. Often it is a big mistake even to start. When the Supreme Court discovered a constitutional right to abortion, Justice Blackmun's opinion tied the right to the first trimester of pregnancy and not the third. That much roughly reflects what most people seem to feel makes sense. There is certainly no economics to this. So far as "external" impacts across the placenta are concerned, they get larger, not smaller as the pregnancy proceeds. So far as the rest of society is concerned, the "external" interests don't vary much over the course of gestation: If the newborn has the genius of Beethoven in his genes, the embryo did, too. Trying to invoke science to resolve the debate, as Justice Blackmun unfortunately did, just muddles it. Fundamental constitutional rights can't depend on the quality of our neonatal incubators.

With pregnancy, as elsewhere, the "internal" gives way to the "external" where society says it does, no sooner or later. Piling on economic mumbo jumbo only confuses things. It is perfectly easy to put food, alcohol, tobacco, contraception, cigarettes, and most portable and seemingly "personal" forms of environmental pollution, on the public side of the line, if we wish to. Cigarettes, the most private form of pollution, are made public through the actuarial processes of public insurance programs, however private the lung cancers remain. The statistics on abortion are easy enough, too. Some quite predictable fraction of the aborted would grow up to thank the Nobel committee for the honor, just as some fraction would grow up to inquire if you want fries with your burger, and still others would end up strapped to a gurney for execution. The trace-toxic models do their statistics much the same way, focusing on the down side rather than the up.

But unless we reckon up all the positive external benefits, all the displacement and marginal effects, all the old hazards undone, the natural hazards warded off; unless we do all that, we are just sketching out the left side of a speculative bell curve rather than the right. With big, obvious externalities, there is no right, and one can get on with fixing them, best of all by establishing property rights. For sufficiently diffuse ones one can't and shouldn't.

It is a normal part of politics, of course, to seek to bottle up the other side's externalities while liberating your own. But it is best to play the game reluctantly, without pleasure or missionary zeal. And it is definitely best not to make up facts about cause and effect, about how all of this is connected to all that, because if one political camp does it the other surely will also. If we are intellectually careless enough, we will end up making the "external" a matter of purely individual taste and prejudice. Each will define for himself the habits he cannot abide in his neighbors, the things that harm him a lot, physically, emotionally, through corrosive effects on his lungs, neighborhood, or spirit. A priesthood of scientists and sociologists will supply expert support to prove how very individual suffering inevitably is, what statistical violence others are known to do to statistical individuals a lot like me. The "externality" is internalized at last. It is whatever I say it is.

The only defense is to remain highly suspicious of broad consequences traced via elaborate models to their ostensible, distant causes, whether traced by zealots of the Left or Right, Soft Greens or Hard ones. Suspicious whether the issue is the murder rate and death penalty, or pornography and rape, or dioxin and cancer. The broader the claim and the longer the chain, the more suspicious we should be. A government generally empowered to hunt down externalities however speculative, however small, however far removed from the ills it is said to cause is one that the Left will end up fearing quite as much as the Right. Beware of the ideologically convenient Big Theory, most of all when it comes on the arm of emaciated facts, however horrifying. It will make civil life impossible.

It is true that externalities are too easily ignored: Pollution remains ubiquitous, nasty, and worse. But it is equally true that externalities are too easily discerned: They have become yet another instrument of the victim culture, in which every individual defines his own environmental poison and demands special protection from it. A society can slouch toward the collective slum and sewer, but it can also slouch toward dilapidation of another kind, in which every citizen believes himself assailed on all sides by invisible evils that no one else discerns. The modest middle course, as Albert Camus advised, is to try to live lucidly in a world where dispersion is the rule.

UNFINISHED BUSINESS

If that admonition does not seem to settle very much, it is because it doesn't. Standards of "clean enough" don't stand still, not in a society that is growing progressively richer. Why should they? What people value keeps evolving; what they will pay to avoid, what they anti-value, keeps evolving, too. Conservative economic policies don't promote stasis, they promote progress. Conservative environmental policies should be progressive, too.

Wilderness conservation and the visibly green must remain at the top of Hard Green priorities. But not to the complete exclusion of all others. Hard Greens know full well that even the invisible can have value, even the innocuous can entail cost, if only because value and cost ultimately lie in the mind of the beholder. People are entitled to dislike chemicals in their drinking water simply because they dislike them, whether the distaste is for fluoride added deliberately by a meddlesome government or tetrachloro-ethylene added negligently by a noisome factory. People are perfectly entitled to prefer pure drinking water, even if contaminants cause them no harm, even if contaminants harden their teeth. People are equally entitled to put up with much less pollution than their parents did. They ought to. Tastes change, as societies grow richer. Sensibilities grow finer. Standards of propriety keep rising. So they should.

The Hard Green policy maker will simply do his best to keep things heading toward property, not away from it, to keep moving from public toward private, from collective prescription toward private control, from government mandate toward market exchange. There was a day when America had no need for fences or titles to land; land was too plentiful to bother with such things. But population grew, the prairies were settled, the Homestead Act was passed, and "natural resource" gradually became real estate. We might instead have handed the prairies over to a Ministry of Lands, to administer from Washington "in the public interest." What we did with the prairies and the Homestead Act a century ago, we are doing with air and water today. It is a vast undertaking. Our initial impulse, unfortunately, has been to affirm public ownership rather than private. But that process is now being reversed, and happily so. Step by step, we are converting environmental goods into forms of property that markets can trade.

If there is one political fact Hard Greens must come to grips with, it is that the stepping will never end, not until people's environmental values stop rising, and there is no reason to expect or to wish that they ever will. The triumph of the Soft Green movement—an indubitable triumph— has been to keep raising those values higher and higher. The Soft Green tragedy has been to raise them up on a flimsy platform of dubious science and meddlesome regulation, regulation that has been inefficient at its occasional best and positively harmful at its much more frequent worst.

Now, at long last, a consensus is slowly emerging as to better means. From Marx to Lennon ("Imagine no possessions"), old-guard Softs, like anti-property pundits in general, have missed the most fundamental point: Property—the idea itself—is ideologically neutral. Yes, property is the capitalist's tool. But the feminist's, too. And the libertarian's. The woman who wants an abortion says it's *her* uterus, not Pat Robertson's. The rancher says it's *his* land, not the marauding wolf's. Your supposed constitutional right to get an abortion, or smoke dope, or ride a Harley without a helmet, all emerged from the Fourth Amendment, written originally to protect "houses, papers, and effects, against unreasonable searches and seizures." Whether you're talking about land, abortion, or environmental protection, you inevitably drift into the rhetoric of property. Pro-choice doesn't mean a thing until you assign title to the real estate. Almost all rational political discourse can be framed as a debate about who should get, or retain, how much title to what. And framing debates that way leads to win-win solutions far more often than any alternative.

With pollution, a consensus is slowly emerging as to means, but no consensus will ever emerge as to ends. Softs will forever be off in the mists, modeling trace effects of this or that, reckoning up costs, and demanding urgent action. And Softs will forever portray Hards as standing in reactionary opposition to environmental progress. Daunting though the political challenge may be, Hard Greens must persuade the public that there is good reason to remain very skeptical about the boundless, misty edges of Soft Green science. Chasing phantoms through those mists is not merely unproductive, it is impoverishing. As I shall argue in the Chapter 9, poverty is not green at all.

For all that, pollution remains a process, not an end, and its abatement will remain an open-ended process, too. The best we can do is stay engaged; pick sensible, tractable, targets; and favor better processes over worse ones. If that is an unsatisfactory way to end the story, it is because there is no end. Our standards of propriety are forever in flux, too, as fluid and changeable as pollution itself. As we judge them today, our grandparents cared far too little about their smoke and sewage, their trash and chemical dumps. Our grandchildren may well conclude the same about us. Let us hope that they do.

9

The Limits to Growth

◆

As pollution reminds us, there are no limits. "Carrying capacity," "sustainability," and similar labels, whether stamped on goods like energy, or bads like wastes, mean nothing. They convey no insight recognized by any testable, verifiable branch of science or economics. The complexity theories of Soft Greens are not science either, they are just doleful hand-wringing about the technologies that give lie to all Malthusian prophecy. No law of geophysics, biology, engineering, or economics decrees: So far, but no farther.

How depressing. Yes, we can fuel our growth from the depths of the planet, instead of from the surface. Yes, we can contain pollution, up to a point, by privatizing it. And yes, we can designate wilderness areas as off limits to economic growth. But how much greener might the planet remain if humanity itself simply stopped growing, if it would just some day declare: Enough, enough, we people have grown enough, let us leave the space that remains for the rest of creation.

Are there really no limits to growth at all? Indeed there are. Notwithstanding all I have said so far, there are, in fact, two. One is poverty. The other is wealth.

POVERTY

Poverty will do the trick, just as Malthus said it would. It certainly did in his day. A woman bore children as fast as she could throughout her short years of fertility, then joined most of her children in the grave. Until recently, that was still the lot of most women in less-developed countries. In the 1960s, when Ehrlich was projecting global famine, the average woman in those countries had to bear six children for the population to hold steady. Until the 1930s, population growth in China was essentially zero, because high mortality offset high fertility.

Pushed far enough, poverty would undoubtedly be green, too. Return humanity to its origins, reduce it once again to tribes of nomadic hunter-gatherers, and great herds of bison will eventually take back the open prairie.

Short of that, however, poverty is not green at all. Despite their small appetites—or could it be *because* of them?—Third World countries manage to generate a lot of garbage, smoke, and trash. They use little, but so very inefficiently that there is tremendous waste. They are frugal and they are destructive, too. They use no pesticides and plow more land, they use no plastics and discard far more organic waste, they eat little meat and shoot more elephants. Behold the impoverished Brazilian farmer, burning down his rain forest to grow a subsistence crop of cassava. He surely cannot be taxed for the species he wipes out; he is living so close to the edge that a tax will wipe him out instead. He is a walking externality. Poverty invariably is.

The happier Third World economies today are what ours were fifty years ago; the unhappier ones are what ours were some centuries earlier. Why should we expect them to be green? We weren't, when we were as poor as they are. Victorian England is not a shining example of environmental rectitude for modern London. Buffalo Bill was not a paladin of wise husbandry on the American range. We have no reason to be proud of our own environmental past, which is, by and large, the Third World's environmental present. Today, half a loaf of the poverty policy means Brazil. It means burning down the rain forest to grow cassava.

POLITICS

Malthus didn't believe government policies were needed to maintain poverty; human fecundity would suffice. But human ingenuity has proved him wrong. Most Soft Greens now recognize that they can't count on laws of nature, still less on capitalism's own internal contradictions, to limit growth. Man has to write the laws and create the contradictions himself. Through his political institutions, he must devise means to curtail innovation, suppress information, increase friction, and promote chaos.

The relentless pursuit of nothing will do. Regulatory crusades against dioxin and PCBs don't reduce any detectable amount of disease, but they assuredly can hobble growth. The costs of such efforts can rise astronomically fast, even as the benefits disappear from sight. The Softs' ever-more elaborate computer models, their speculative prognostications about the far future, their preoccupation with eco-phenomena only they can discern, their ant-heap sandpile theories, their efficiency proclamations, their redesigning of toilet-cisterns, their recycling of potato peels, their determination to pursue every last particle and molecule of modern industrial life because it might affect female hormones or male sperm counts or global temperature: All of these things sow economic confusion, postpone investment, waste time, disrupt markets, hamstring technology—and so, limit growth. Such policies have nothing to do with saving the environment from humanity's wastes. They generate waste itself: economic waste.

When it comes to approving new enterprise, a fractured, labyrinthine bureaucracy is best. Place uranium under one agency, coal under another, hydroelectric under yet another. The objective is not to choose intelligently among imperfect alternatives, it is to obstruct new growth, wherever we can. Oppose oil for its scarcity, coal for its pollution, and uranium for its complexity. The *Exxon Valdez*, acid rain, Three Mile Island: Each is a growth-impeding opportunity to be seized and exploited. The most important battle is ideological. Infiltrate the educational agenda of the public school. Fund research, and more research, channeled of course toward the correct, trans-scientific conclusions.

That was how Soft Green got started in the 1970s. Rather than unleash markets to overcome scarcity, Western governments enlisted bureaucrats to institutionalize it. They resolved to limit population, limit consumption, limit growth, by welding shut the valve of high technology. It was the Softs' happiest decade. By such means, albeit with the help of Arab sheiks, Softs managed to limit growth quite a lot. Earnings fell, wealth fell, and yes, for a time, consumption fell, too.

In a democratic society that attends to vocal minorities, it is not too hard to thwart any single, well-targeted technology or enterprise, even a large corporate one. But there is no method to the process of obstruction, there is only Soft Green madness. Small wonder, then, that the environmental consequences are dreadful. The power plant operator nurses along his old coal-fired, nuclear, and hydroelectric facilities, just as a capital-short teenager nurses along the ancient Buick surrendered by his parents. The capitalist stops staking billions on an opaque regulatory future. When the capitalist finally gets around to buying, he opts for another basic Buick: the old and familiar design, the conventional fuel, the established pollution-control technology. The best-traveled regulatory path is the smoothest. Opt for technologies that rely less on capital, more on fuel and labor, just-in-time costs, that can be passed through to ratepayers the fastest.

Believing they really could halt growth in its tracks, the Softs of the 1970s made a hugely risky environmental bet, and they lost. They undertook to stop growth by throttling down the biggest, most efficient, and cleanest engines of capitalism. Nuclear power first, then the rest of big, hard electricity. But when the final tally was in, the Softs hadn't stopped growth, after all, not of electric power and not of anything else. They had merely thwarted, here and there, the most clean, efficient, and economical engines of growth. Nuclear, the cleanest of them all, had been brought the lowest.

Looking back, the honest Soft must get a terribly queasy feeling in the pit of his stomach, the kind you get when you've worked desperately hard to get somewhere—swim across a lake, say—but you discover about half way that you can't make it, and you realize that half way isn't half a victory, it's a disaster. That's what the Softs have achieved with energy since they set to it in the 1970s. Our total energy consumption has risen 20 quads in the twenty years since they began to work their mischief. Among fossil fuels, coal has accounted for most of the growth. And greener though it is than Soft alternatives, coal is indubitably the least green among Hard ones. The Softs were so busy fighting nuclear because the atom bomb scared them and fighting oil because the Arabs did, that they forgot about the West Virginia coal miner, who did not scare anybody much at all. Lovins preached biomass and in the end, we went for it: the biomass most readily at hand, the kind long buried in the Appalachians and the foothills of the Rockies.

EFFICIENCY

Thwarted in their pursuit of real poverty, the Softs next embraced the fake kind. Efficiency, like the Diet Coke, would let us be poor and rich, too, both at the same time. We'd be stingy at the fuel pump, but still profligate in cruising down that ribbon of highway. Frugal at the back end of our refrigerator, but the cup still overfloweth with ice out front.

Now, after a couple of decades of prescribing efficiency standards willy nilly, the Softs find themselves eager to report that efficiency has worked, that growth has indeed been tamed. But the numbers are squarely against them. Year by year, national consumption of energy still rises, with no perceptible effect from all that prescribed efficiency.

At this point, Softs split into two camps. Casual Softs—the great majority, it seems—just don't dwell on real numbers. Or else they fake them. Their favorite trick: Report new "growth" figures *per capita*. But per capita reductions in smoke and sewage don't clean the air or water; per capita reductions in shooting don't save the bison or the eagle; and per capita reductions in energy consumption don't reduce energy consumption. Doing your Malthus *per capita* is oxymoronic: The entire Malthusian argument centers on capita, on growth in the aggregate. And since the dawn of the Industrial Revolution, a big part of the growth has taken the form of more people, not more wealth. The Casual Softs are like the chronically obese clutching at their diet sodas. They keep insisting that the diet is really working, that it will halt growth, or even reverse it, quite soon. So much saving, from all this efficiency we've prescribed, simply could not permit such relentless bloat. The scale must be broken.

But it isn't, and serious Softs know it. They know how to measure efficiency gains, which are considerable. They also know how to count new power plants and oil tankers. They know how to read the scale, and they hate what it reveals. They make no promises about painless, diet-soda weight loss; they demand instead a real diet. They know efficiency buys them nothing. They do not want resources used efficiently; they want fewer resources used, quite a different thing.

I shall continue with growth shortly, but let me finish burying efficiency first. All other things being equal, with both population and individual appetites held constant, what are efficiency's true environmental consequences, anyway? It depends. Efficiency does for the environment exactly what it does for the economy as a whole. True efficiency improves it. False efficiency degrades it. And Softs don't know the difference. By imposing engineering efficiency on a marketplace that wouldn't otherwise buy it, they promote economic *inefficiency*. And by doing that, they don't promote green, they suppress it.

Outrageous though the proposition may sound, it is the free market that delivers real efficiency—economic efficiency—in refrigerators, cars, and homes. Not because it minimizes use of oil, watts, or any other single factor of production. But because economic freedom lets us optimize our economic choices along the many different dimensions that value and wealth comprise: comfort, pleasure, and satisfaction. Yes, a gas-guzzling sports utility vehicle is, indeed, "efficient." It wastes gas, but it is still efficient in the holistic, economic sense, at least once pollution costs are properly internalized. The gas guzzler makes the driver who buys it richer, in that he is happier owning the car than something else he might have bought with the money. That's why he bought it. That's what "efficiency" means, in economic parlance. And as I shall argue in the following section, *that* kind of efficiency, economic efficiency, is the only kind that limits growth. It does so by making us richer.

But the Softs, as I say, just don't distinguish economic efficiency, which in the end *is* green, from the thermodynamic kind, which isn't, at least not the way they set about delivering it. They deliver it by prescription. And whatever it may do for the motor in your refrigerator, prescription is economically inefficient. Economic efficiency is never conjured out of the depths of government codes, registers, proclamations, prescription, regulation, and meddlesome forms of tax or subsidy. However loudly they may be rationalized in efficiency's good name, those things deliver only its counterfeit, its precise opposite. Governments can't prescribe or impose efficiency, not the economic kind. Free choice in a free market is as efficient as you can get.

We ought to have learned that by now. "Efficiency" planners have been around a lot longer than Soft Greens. Socialism, recall, was "scientific." It wasn't just going to make cars or refrigerators efficient, it was going to make whole economies efficient. It was going to wring the waste out of capitalism itself. Yet centrally planned industrialism, for all its desperate pursuit of efficient production, produced far fewer amps and ingots than the honestly efficient capitalists. Everywhere they came to power, central planners laid waste.

Economic waste, and environmental waste, too. They despoiled the environment with gross, arrogant, blundering, callous, stupid savagery almost unimaginable to us capitalists. Murray Feshbach and Alfred Friendly, Jr., set out some of the appalling details in a 1992 book, *Ecocide in the USSR*.[1] For seventy-five years, the Commies systematically poisoned the air, soil, and water of one-sixth of the Earth's land mass, stretching from Poland to the Pacific. Draining the Aral Sea in Central Asia to irrigate cotton fields, they precipitated the "greatest single, man-made ecological catastrophe in history." In Western Siberia they poisoned the Tom River and the Angara River, and hence Lake Baikal, the Enisel, and the waters northward to the

Arctic Ocean. In European Russia they poisoned the Volga, Dnieper, and Don. In the steel city of Magnitogorsk their open hearth furnaces dumped 870,000 tons of atmospheric pollutants per year into the air. "No other great industrial civilization so systematically and so long poisoned its land, air, water and people."

None of this happened, not officially. The Communists had enacted some of the strongest environmental laws in the world. Unfortunately, they neglected to obey any of them. Capitalists have something to lose under the law: their own wealth. Public bureaucrats don't. Capitalists are said to sacrifice nature to material abundance. The Communists sacrificed nature far faster and more ruthlessly, with no abundance to show for it. Capitalists tend not to feel heroic; they merely aspire to conquer Wall Street. The Communist agitprop depicted a heroic society that would conquer all of nature. And because everyone owned the land no one took responsibility for it.

Oh yes, to be sure, those were the *old* efficiency planners, red socialists, not green ones. The Soft Green has seen the *new* "future that works." It's no longer the system that will generate more electricity than any other; now, it's the system that will generate less. Because we will use it so efficiently. Soft Green is for the 1990s what the Dnieper Dam and the newest red salmon-canning factory were for the 1930s. Economic efficiency then, environmental efficiency now, the laws of economics then, the laws of ecology now, the oppression of workers then, the oppression of all life in the biosphere now. Soft Greens will suppress demand for electricity as ruthlessly as the Reds once promoted its supply. They aren't going to build an enormous salmon-canning plant; they are going to package millions of tons of tofu instead and teach us all to like it.

But they won't, of course. Central planning is never efficient, it is congenitally stupid, whether aimed at electricity or salmon, supply or demand, economy or ecology. Central planners disdain, and so never engage, the vast reservoir of initiative and intelligence in ordinary people. Even while they remain honest and well intentioned, which they rarely do for long, central planners get hopelessly muddled. They want to manage the scarcity wisely, but they also want to be fair. So they build big dams to supply new power, then sell it so cheap it is used in profligate excess. They mandate clean cars, but won't touch clunkers driven by the poor, the worst polluters. They prescribe the size of our toilet cisterns, but they won't put a price on water; think of the poor. They slap a gas-guzzler tax on the rarely driven Rolls, but never on a million-mile Chevy.

They demand clean air, and they demand jobs for coal miners. However imminent the famine, fairness to the fecund populace can never mean raising the price of food. Central planning invariably ends with huge bureaucracies laboring to conserve salmon in a toilet cistern.

Central planners are the only ones who can turn the green triumph of Hard technology into environmental ruin. In the hands of a capitalist, the central electric power plant is the greenest option there is. In the hands of the central planners it is Chernobyl: not hard power gone wrong, but its management gone wrong. Central planning always goes wrong. The Communists built and operated power plants fanatically, because for them, big power was a symbol of political triumph. Why should we be surprised they turned Hard power into a green disaster? They turned all wealth into poverty.

If there is one political lesson we should not have to relive and relearn, it is that central planners always promise efficiency and always deliver its opposite. When workers were poor, they promised efficient wealth and delivered three generations of waste. Now that environmental poverty looms, they promise efficient green. They use the word freely enough, but they simply don't know what efficiency means. Out of ignorance that deep there comes only one thing: despoliation and decay.

It is the free market that is efficient: spontaneously efficient. The Hard technology of modern capitalism is fantastically efficient. It produces food, power, transportation—everything—far more cheaply and abundantly than Soft alternatives. It solves every problem except the problem of wealth. Hard technology does not reduce consumption, it increases it. It makes us richer, not poorer. Which means it lets us grow and grow, if we choose to.

That is what ultimately damns Hard technology in the mind of the Soft Green. Not its inefficiency or instability, but just the opposite: its relentlessly stable efficiency. Hard technology is so efficient, so stable, that it makes us rich. Which—the Soft Green believes—means growth, "the creed of the cancer cell."

WEALTH

The outside world imposes no limits to growth on a society that unleashes the real kind of efficiency, economic efficiency. Resources don't limit growth; markets find or create new ones. Pollution need not limit growth; turn pollution into property, and capitalists will package pollution and transform it into wealth. The catastrophic collapse of sandpiles won't limit growth; capitalists learn how to design technological honey. On all sides, free markets create abundance. Efficiency—the real kind, discovered by markets, not bureaucrats—creates still more abundance. Complexity creates efficiency, which creates still more abundance. Yes, Hard technology is so efficient, so stable, that it makes us rich.

And what follows from that is exactly what Malthus predicted, with just one critical correction. First, the part that Malthus got exactly right. Free markets create wealth, and wealth propels life. Richer people live longer. More women grow to adulthood, to bear children of their own. Population rises.

That much of the proposition has been confirmed empirically. Free markets took control of economic life in the west at about the same time as Malthus was publishing his pamphlet. In the two centuries since, the west has grown steadily more wealthy, and mortality has dropped apace. Population has expanded to consume the land (or, more generally, resources) available. Just as Malthus said it would.

With, as I said, one critical correction. Affluence increases longevity. But it also lowers fertility. The data on this point are equally unambiguous. Developed-world fertility has been falling quite steadily for two centuries. In the United States, it dropped from eight children per woman to two. Fertility has fallen the most in this century, as affluence has risen the most. In what the United Nations calls the "More Developed Regions," the "total fertility rate" (roughly speaking the average number of children born per woman) has fallen from 2.8 children per woman in the 1950–1955 time frame to 1.6 at the end of the century. That puts it well below the replacement rate.

Why? As Ben Wattenberg has pointed out, "Young men and women conceiving children, or not, often in a darkened bedroom, aren't thinking about pretty charts or about an invisible line called 'replacement.' They are thinking about a good life for themselves, in quite new, modern circumstances." Ordinary people don't know the statistics but they have their instincts and their eyes. They generally have a pretty good idea what prospects their own children will face. Roughly speaking, if the risk of dying by age thirty is 50 percent, the average woman will have to bear over 4 children simply to hold her own in the race for genetic survival. In the United States today, where the mortality figures are far lower, 2.08 children per woman will do it.

Affluence allows parents to raise fewer, more robust children. Producing food abundantly, in other words, is a highly effective way to limit population. This is the critical thing Malthus missed. The generations since Malthus have not simply contrived to consume every last ounce of new resource that free markets produced. They have contrived merely to hold their own.

So why has the population of developed countries continued to rise? Simply because people did not commit mass suicide when they got richer. Population is determined by the combined effects of fertility and mortality. And for obvious reasons, changes in mortality and fertility take several generations to shake out fully as net changes in population. For most of the last two centuries, mortality rates were dropping faster than fertility rates, so population grew fast. But by the twentieth century, the "several-generation" clock had run. First, mortality and fertility came into balance. Then, in the latter decades of the century, declining fertility overtook declining mortality. Putting aside effects of immigration (a zero-sum game), populations in the developed world have now stabilized. They will soon begin to shrink.

Wealth is beginning to replace poverty as the force that limits population growth in developing countries, too. In recent times agriculture and technology imported from the west have brought the developing world a growing measure of affluence, and affluence has lowered fertility. The fertility rate in the U.N.'s "Less Developed Countries" has fallen from 6 children per woman in the 1960s to 3 today and continues to fall fast. The rate in India today is lower than the American rate in the 1950s. Fertility rates in most sub-Saharan African nations have dropped from 6.5 to 5.8 and are falling steadily.

For the planet as a whole, population continues to rise, because of the several-generation lags between fertility and mortality changes on the one hand and population changes on the other. World population was 1.6 billion at the beginning of the century and 2.5 billion in 1950. It is about 6 billion at the end of the century and is projected to reach somewhere between 7.5 and 9.4 billion by 2050. Then it will start shrinking. Down to about 6 billion in 2100 and 4.3 billion in 2150, according to long-range U.N. projections. As Wattenberg notes, "a bust, like an explosion, moves in a geometric progression."

RICH MAN, POOR MAN

As the Soft Green sees it, personal wealth—which he invariably confuses with macro-economic growth—is the antithesis of green. To him, more wealth can only mean more growth, more aggregate consumption, and thus a larger, heavier human footprint on the face of the globe. But he is wrong, flat wrong. He makes two fundamental mistakes, both rooted in his bankrupt political ideology.

First, like Marx, he reviles high technology, which keeps affirming the power of markets and capitalists and which keeps repudiating his predictions of scarcity. But high technology is what allows us to grow bigger and greener, both. It lets us find our wealth away from the land—literally *off* the land—by prospecting below and above it instead. It lets us live in the third dimension, where the rest of life isn't. Nuclear power, the capitalist's most despised invention, uses a truly minuscule quantity of lifeless mineral, contained securely on a minuscule amount of real estate, to power a metropolis. The military uses it to propel a submarine or aircraft carrier around the globe and to power its satellites for decades, precisely because so little material provides so much power, a vital advantage in times of war where supply lines are unreliable. Biotechnology, equally despised, supplies comparably great gains in agriculture, the one system for capturing solar power that is in actual, widespread use today. Free markets spur innovation, and innovation lets us loose the surly bonds of Earth. The limits exist only in the mind.

The Soft Green's second big mistake is even more fundamental. His mind-set is macro, not micro. He thinks of growth the way he thinks of everything, in collective terms, as a Marxist would: more total pig iron, more megawatts, more national population. His metrics are the traditional Malthusian ones: Nation States, World Models, Global Trends. He doesn't see the world through the eyes of the individual.

The individual, however, doesn't have a GNP, he has a bank account, which he wants to grow. He doesn't have a population trend, he has a family, which he wants to raise. Yes of course, the Soft Green replies, but the whole is just the sum of the parts. Multiply such aspirations across the population as a whole, and you are right back to more people, more consumption, the heavier footprint.

But you aren't. Private wealth does not work that way at all. The individual's desire for more wealth has no limits, but his appetite for hamburgers does. Up to a point he will eat more, drive more, build more. Then his basic Malthusian desires are satiated. The individual does not wish to seize and eat, to seize and drink, to seize and fornicate, more and more, without limit. Happily, he just isn't built that way. The individual desire to grow within, to grow his waistline, his family, his estate: *that* is what eventually reaches a limit. The real limits to growth are rooted in nothing more complicated than that.

In the West, obesity is a disease of the poor, not of the rich. Worldwide, tobacco is the personal pollution of the poor, not of the rich. Television is the mental opiate of the poor, not of the rich. Waste and inefficiency are the lot of the poor, not of the rich. People struggling to survive don't much care about nature, except where they actively fear it. It is wealth that distances such tastes and tendencies. And when the rich man reaches the private limits of his own consumption, he puts his wealth into other things. That is what people always do as they get richer.

First, into better lives for his children and grandchildren. And the better he can take care of them, the fewer he has. The poor reproduce as Malthus said they would, as fast as they can, and pay a fearful price in the high mortality and dismal prospects of their offspring. They are "inefficient" here, too, if such a coldhearted word can be used to describe child mortality. The rich man secures his genetic posterity through quality, not quantity.

And with that secured, he pours his wealth into green. He puts it into land and then jealously conserves the estate, fighting off encroachment by the fecund poor. He craves open space and buys it as soon as he can afford it, in his backyard, his neighborhood, his state park, and through his elected government, too; that part of his tax he generally pays quite willingly. He still wants to get richer, but he counts his wealth the way a rich man does, not by food in the larder, but by the quality of life that surrounds him. He puts some of his wealth into charity, some into art, some into his bird feeder, and some into the prairies beyond. The richer he gets, the farther the footprint of his wealth must extend. It extends first to helping neighbors, then to conserving nature. It has nowhere else to go.

The poor man doesn't behave that way at all. We didn't either, when we were still poor. The feudal lords of England once conserved their forests for the pleasure of hunting. But the peasants who vastly outnumbered them scraped and gouged little scraps of land, poached on the lord's preserve if they dared, and were hanged when they were caught. Of course they gouged and poached; their alternative was to starve. Wealth impels us to control our own Malthusian proclivities effectively and humanely; poverty impels us only to surrender to them and end up in misery, as Malthus said we would. Wealth solves the problem of scarcity with abundance. It solves the problem of population by defending life, not by surrendering to death. Ordinary people, rich or poor, instinctively know that.

So how much wealth does it take to make a man green? Just when will he stop diverting his wealth into a bigger powerboat for racing through the Everglades and begin putting some of it into the Everglades themselves? It depends ultimately on where he finds his pleasures. Some find their personal needs easily satisfied; others don't. There are hermits and there are hedonists. For one and all, however, "conservation" has nothing to do with the size of your car or refrigerator and everything to do with the size and character of your appetites. The green path is not one that "conserves" fossil fuel or "invests" in solar power; it is one that conserves land, nature, and environment and invests in the same. The true "conservative" is efficient, which makes him richer. He is frugal, too, frugal in his own aggregate consumption. He is profligate only in his determination to accumulate his wealth in the form of unspoiled land and untamed nature. The only form of poverty that is green is the poverty chosen freely by the man so rich— so rich in capital, so rich in spirit—that he willingly chooses to spend his treasure preserving and advancing more life than his own.

Experience teaches that it is far easier for the rich than for the poor to pass through the eye of the green needle. The anxiety and stress of chronic poverty do not make people green. Green is what people become when they feel personally secure, when their own appetites have been satisfied, when they do not fear for the future, or for their own survival, or their children's. It is wealth itself that gives most ordinary men the confidence to be generous to the world beyond. It is the rich who can afford poverty, so to speak; it is they who can declare, enough, enough, I need no more, it is now time for me to grow things other than myself. It is the rich who can be thin because they know they will always have plenty to eat. It is the rich who can be green because they no longer have to choose between their own survival and nature's. It is when you are rich enough— and only then—that your very conception of wealth can expand to encompass not just the personal but the public, too. It is only when you are really rich, in body and spirit, that all forms of wealth, personal, social, and environmental, converge.

Of course the rich consume more than the poor. "Rich" inevitably entails some of that. As it happens, we also produce our abundance far more efficiently than they produce their scarcity, and that makes us richer still. The peasant hunched over his cow-dung fire is not efficient, not green, not living in happy harmony with nature. He is just poor.

And of course the rich generate more waste than the poor. What goes in must come out; productions of goods and garbage rise together, even if the efficiency of our production lines makes our balance more favorable than inefficiency makes theirs. But wealth also lets us handle our waste better. Nobody who knows Calcutta, Mexico City, or tribal Africa could possibly doubt that. The desperately poor live in their own compost heaps and septic systems, and there is nothing green or environmentally harmonious about that, except that it keeps mortality rates appallingly high. Flush toilets and closed sewers make our immediate, human environment far more pleasant than theirs. Sanitary landfills, built to modern standards, take care of the environment at large, closing the carbon loop, returning much of it to the earth whence it came.

And without doubt, the rich are more efficient, in the engineering sense of the word, than the poor. Efficiency is what accounts for much of our wealth. The rich capitalist microscopically tunes and adjusts the fuel-air mixture in his giant power plant because each tiny uptick in efficiency saves him millions. The peasant hunched over his burning cow dung does his best, too, with the means at his disposal. But he has no means. He wastes fuel, pollutes, and fouls his own lungs as he tends his tiny flame. The capitalist is richer, consumes far more, but consumes far more efficiently, too, because he feeds his cheap fuel into such capital-intensive, meticulously designed, complex technology.

The notion that our wealth derives from—worse still, *causes*—their poverty is arrant nonsense. Western wealth does not cause Third World poverty, neither the poverty of the people nor the poverty of the land from which they scrape their meager existence. The overfed American farmer grew the grain that fed the peasants in Biafra and Bangladesh during the years when Ehrlich predicted they would starve. The overfed American agronomist then supplied them with the high-yield, pest-resistant crops that now help them to feed themselves. As for the land, poor countries are horribly bad at conservation because they can't afford it. Only the rich eco-tourist or trophy hunter can persuade them otherwise. We arrive, on fuel-guzzling planes, with our capacious guts poured into our comfort-fit jeans and then—not a moment earlier—the protecting of the elephants begins. Consumption always comes first, conservation second, for the simple reason that without consumption first there is no second. However much green theorists may insist otherwise, conservation is not a necessity, it is a luxury. Just ask anybody who is really poor.

With whales, elephants, and rain forest, it is Western wealth that has propelled every last tusk of conservation. For the poor, the elephant is a mountain of meat, the whale is a barrel of oil, and the rain forest is a place to grow cassava once the monkeys have been shot and the undergrowth cleared by fire. The rich have learned to live in three dimensions and are thus retreating from the land. While the poor burn down their forest, we regenerate ours. We have solved the problem of population, too; our capitalist wealth has solved it. If the poor are now making rapid progress on that front, too, it is by building wealth with Western know-how.

Nobody who studies the reality of conservation can seriously doubt any of this. The rich, not the poor, are the ones actively committed to conserving wildlife, forest, seashore, and ocean. The charge that the rich are the despoilers, the exhausters, the expropriators of the planet's biological wealth is simply false. Rich people did not invent slavery, they ended it. Rich people did not invent the oppression of women, they ended it. Rich people did not invent environmental destruction. The rich are ending it.

Malthus got half the story right. Resources do affect population. Poverty limits it, by letting people die. But wealth limits population, too, by letting people live. It is by mastering the problem of poverty that we master the problem of population. And it is by mastering the problem of poverty that we master the problem of growth. Growth is not "the creed of the cancer cell." It is the creed of liberated women, a stable population, and a better life. Among humans, the famine need not come. Among humans, wealth limits fecundity far better than poverty. We do not need to starve population into its limits, not by exhausting our land and not by political means either. Malthus was half right. Poverty is indeed one solution to the problem of scarcity. Wealth is the other. Poverty is indeed one of the limits to growth. Wealth is the other.

CAPITALIST GREEN

Few people still believe that top-down prescription can deliver economic wealth. A dispiriting number still believe it can deliver the green kind. Capitalism, we are still instructed, is a voracious, dirty system, as "congenitally blind" to environmental interests as it once was to labor's. So says the *Encyclopaedia Britannica*.

The political prescription hasn't changed much either: It's still less market, more government. Replace the Invisible Hand with new management that is more frugal, stable, efficient, fair, democratic, and now, of course, green. A government board will maintain the equilibrium that markets cannot supply. It will boost supply, of good things like windmills or solar power, and ration consumption of bad ones, like oil. It will slow growth from the top down, by dictate, by treaty, by mandate, from the World Bank, from Rio, from Kyoto. We will now deliver green by the same political means that were going to deliver pig iron.

But those means always fail, because they are at war with human nature. One cannot impose altruism and beneficence from the top. Not efficiency, either. Trying to do so lowers economic efficiency, which lowers wealth, which undermines green. Efficiency makes us greener only when it makes us richer. It *does* make us richer, but only when it is adopted by choice, not compulsion, by the capitalist, not the bureaucrat, by the market, not the government. Forcing people to buy fuel efficiency they don't want, or power savings they wouldn't otherwise bother with, makes them poorer and so, less green. Allowing people to pursue efficiency by choice, as free markets do, makes them richer and so, greener. The pursuit of efficiency as its own green end is a distraction. And a harmful one—because it is an impoverishing one—when it becomes coercive.

Free markets are green because they are efficient in the economic sense of the word, the one that matters. They promote efficiency by promoting the accumulation of private capital, the greenest factor of production. A campfire is cheap, inefficient, and dirty. A central power plant is expensive, efficient, and clean. Meticulous engineering design and advanced technology substitute for a lot of fuel, a lot of surface. Private capital is the best substitute for every other resource. It is private capital that makes it possible for us to extend our reach into the third dimension and so minimize the imprint we leave on the surface.

The central planners understood the magic of capital, too: They were, indeed, profligate investors in massive dams, power plants, and factories. If building industrial monuments were the whole story, they would have been the greenest of all. But to work green magic, capital has to be generated, preserved, and managed efficiently. The central planners knew only how to waste it. For all their grandiose capital spending, they created less capital, invested less capital, and conserved less capital. They wasted their capital, and they wasted their environment, too.

The only substitute for capital itself is human intellect. The mystery isn't why we consume so many resources so fast, from so deep in the Earth. It's why thousands of generations of shivering, starving humanity left so much wealth untouched. And it isn't much of a mystery, at that. Oil two miles beneath Alaskan ice or Saudi sand is not wealth at all. It doesn't belong to anyone, least of all to "the world." Nor do minerals on the sea bed or on the surface of the moon. We call such things "resources" by convention, but the "resource" is not the stuff itself, it is the know-how for getting it. Uranium and coal are not wealth worth "conserving"; they are things long dead and buried that today's technology, for the first time in human history, can use as a substitute for land and life on the surface. Oil is not wealth, and it is not biosphere; it is what lets us conserve biosphere. Nobody even had the slightest idea how to get it until Colonel Drake thought to drill the earth rather than harpoon the whale.

But capitalists devour their own future, Soft Greens protest. To the contrary, capitalists discover it. In a free market, every individual chooses continuously between consumption and saving, between bird in the hand and bird in the bush, between what we know how to do today and what we can reasonably expect to be able to do tomorrow. In the aggregate, individuals make far better choices from the bottom up than the wisest planners can impose from the top down. So much wiser that scarcity never arrives.

But it will, the Soft Green insists. Our *Limits* models say so. Our sandpile models say so. Yes, central planners always say such things. They are always telling us that the future—the one they discern—needs something different from the balance the individual strikes between today and tomorrow. Why should we believe the central planners when they now insist that saving green resources is everything? Half a century ago, they were saying exactly the opposite, squandering the green for the sake of current development and consumption. Central planners never get the time value of money right. They never get the value of anything right. Only free people and free markets come anywhere close.

So long as central planners are in primary control, the environment will never be secure. The people who couldn't deliver the electricity or the canned salmon won't deliver cool air or oceans teeming with fish either. The Communists all acted in the name of progress as they laid waste to their vast lands. All their intentions were officially good. It was their system that was bad. In 1990, seventy Soviet factory directors were fined for discharging polluted water. Their fine was 50 rubles, the price of two packs of imported cigarettes. Soviet economic theory put no price on resources held by the state, and the state held everything. Leaks in some oil well pipelines spilled as much as 10 percent of the nation's production. No money-grubbing capitalist would ever have allowed *that* to happen.

The one permanent scarcity is scarcity of knowledge. And central planners squander and dissipate that, too, just as they squander pig iron and

forests. For all their economic models and theories, for all their endless cal-
culations about the future, the main business of all central planners is to
suppress information, the information the free markets elicits, accumu-
lates, and conveys far better than any alternative. Beyond wealth, and beyond green, free markets produce knowledge, the most precious resource
of all, the one that allays scarcity of every other kind. Green scarcities included. It is the continuous accumulation of intellectual capital that allows
us to generate more wealth and still more knowledge, to grow more food
on less land, to capture more energy from places ever farther removed
from the thin film of life that dwells on the surface of this planet Earth.

We are wealthy enough today to leave much to our children, and we
should. What we owe posterity is the productive inventiveness of our
own times: better grains, superior plants, better engines and power
plants. We owe posterity freedom: free markets, fertile minds, and plain
old wealth. It is wealth, nothing less, that will let us, and our children,
too, conserve the intellectual capital of the biosphere, the most ancient,
varied, and rich manuscript ever written, the genetic code, the Rosetta
stone of life itself.

THE CONSERVATIVE COMMUNE

How then can Hard Greens still defend any corner of the conservative
commune? Why should we ever favor public parks, forests, marshes, or
shores? As I discussed in Chapter 6, we favor them where we must, favor
them gladly where there is no other effective way to create or conserve
the values at hand.

This is hardly a novel proposition for conservatives. Affirming and
protecting our liberties is a civic duty, and an armed citizenry can play a
role there, but it is also a duty we confide to the Joint Chiefs of Staff. Private museums, galleries, and archives do much to preserve our culture,
but the Smithsonian is there, too, and few conservatives would arrange
things otherwise. Most of our scientific and technological research is
conducted under private auspices, but government has played a legitimate and important role, too, in funding basic research and exploring the
frontiers of space. Public authorities play a vital role in conserving bat-
tlefields, monuments, and historical buildings, because they must, because some valuable places and things cannot practically be conserved in
any other way. Green conservation is no different; most of the initiative
must be private, but some important part is inescapably public.

Some goods require public ownership simply because they are too large or fluid to be encircled by private fences. With others, public ownership enhances the value. Conservatives understand that a nation needs a flag, an anthem, a pledge of allegiance, a day of Thanksgiving, and with each of these the value inheres in the public character of the good. Some values exist, or are greatly enhanced, *because* they are public, *because* they have been defined as part of the national patrimony.

What central planners never grasp is that most values are conserved only so long as they remain securely private. The Soviets tried to make pig iron and megawatts into symbols of national greatness, too. They thought that every last bolt and barrel and loaf of production would glorify the state and that the glorious state would, in turn, make all production more efficient, more rational, more scientific. They denied all lines between public and private. As we have seen, so do the Softs. For all practical purposes, they would take us right back to the 1930s, where every last vestige of production and consumption, every power plant, gas furnace, and barbecue, every pound of carbon and chlorine, every trace of dioxin and estrogen, would require government prescription. All that has changed is the official statement of objectives at the top of the page: economical then, ecological now.

As conservators of green communes, Hard Greens won't repeat that mistake. We recognize separation of powers as the key. The objective of conservation isn't economic, it is anti-economic. The locus of conservation isn't in every school, home, and toilet cistern, it's in designated wilderness areas—large ones, we hope—that we can draw on a map. The instruments of conservation aren't propaganda and the indoctrination of school children; they are the political designation of particular conservancies; delineated lands, lakes, and shores; and the creation of new rights of private property to cabin major pollutants.

It is by affirming the legitimate government role in the truly public sphere that we can be all the more determined to exclude government more strictly from the private. We affirm our determination to conserve Yellowstone and its richly varied wildlife for our collective posterity; we also affirm the right of the private landowner to protect his home from brush fires, even if doing so endangers a rat.

Yes, the private sphere must come first: private land, private control, private wealth. And happily, that is where most of the world is now headed. The main trouble for central planners, red or green, is that most ordinary people aren't buying their claptrap any more. Free people prefer wealth to poverty, and free people invariably find it. Around the globe, free people want green, but not poor green. They want rich green. Hard Green. Capitalist green.

And the most responsible thing we can do is to see to it that they get it. There are two limits to growth, wealth and poverty. Capitalism is the road to wealth; central planning, the road to poverty. Wealth is Hard, but it is also green. Poverty is Soft and not green at all.

10

Ethics in
the Green Lifeboat

◆

"Here we sit, say 50 people in a lifeboat. To be generous, let us assume our boat has a capacity of 10 more, making 60." And practical minds will not be so "generous": The latter estimate is possible only if we "violate the engineering principle of the 'safety factor.'" The boat certainly could not hold 100, say. And yet new souls keep clawing their way aboard. Painful though it may be, the ethical course of action is obvious. Some must be left to drown, that the rest may live.

And so it is, our essayist declares, with the green lifeboat of "spaceship Earth." We occupy a small vessel of "limited carrying capacity . . . already exceeded." "The exact limit is a matter for argument, but the energy crunch is convincing more people every day that we have already exceeded the carrying capacity of the land." No amount of innovation or human ingenuity can alter that reality. Though a "splendid technological achievement," the green revolution will backfire in the end. The oceans will soon be exhausted too, and "[n]o technological invention can prevent this fate."

In such circumstances, the sharing of loaves, fish, and such is no longer ethical. "The fundamental error of the sharing ethics is that it leads to the tragedy of the commons." Fisheries and ranges and pollution must be lifted out of the ambit of the commons, or ruin is inevitable. We must limit immigration. Curb reproduction. Let hunger run its course. A world food bank "creates an unacknowledged commons" and "can be relied upon to make matters worse." "Every life saved this year in a poor country diminishes the quality of life for subsequent generations."

Who is this misanthrope who would stop us from feeding the hungry? An armed survivalist holed up in some cabin in Idaho? A crank handing out smudgy pamphlets at a convention of the NRA? No, it is Garrett Hardin, professor of human ecology at the University of California at Santa Barbara, widely renowned sage and ecologist, author of the enormously influential "The Tragedy of the Commons." Hardin's 1974 essay, "Living on a Lifeboat," is almost equally influential.

The metaphor still floats today, although lifeboat Earth is much more crowded than it was a quarter-century ago. Pentti Linkola, an amateur biologist, eco-fascist, and one of Finland's most celebrated authors, knows that the Earth's population is already more than double what the planet can support. The west must end all aid to refugees and the Third World. Abortion should be mandatory for women who have already borne two children. We occupy a sinking ship with one hundred passengers, and a lifeboat for only ten. "Those who hate life try to pull more people on board and drown everybody. Those who love and respect life use axes to chop off the extra hands hanging on the gunwale."[1]

ETHICS IN THE SHADOW OF ECOLOGICAL HELL

To a Hindu, killing a cow is an even greater sacrilege. In times of famine, Hindu children once starved amidst hundreds of pounds of edible bovine calories. How can a just society put a cow's life above a child's? Marvin Harris sets out a model ecological answer in *Cows, Pigs, Wars and Witches*.* The taboo sustains countless millions of rural Indians in intricate interdependence with their animals, land, soil, and climate.

The main economic function of the zebu cow is to breed the oxen that do most of the plowing on Indian farms. Milk is an essential food; manure supplies fertilizer, fuel, and a substitute for plaster. And the cows are scavengers, feeding mostly on grass, stubble, garbage, and the inedible by-products of human food crops. Even a dry and barren cow is thus an essential economic asset. But the temptations to slaughter and eat your cattle are as frequent as the consequences are disastrous, because of sharp year-to-year variations in Indian harvests. Slaughter a cow in time of

---

*Marvin Harris, *Cows, Pigs, Wars and Witches: The Riddles of Culture* (New York: Random House, 1974).

drought and the farmer slaughters his own economic posterity. Killing cows is an even greater threat to that society's long-term survival than famine itself.

Substitute "nature" for "cows," and Harris turns into Hardin. As I suggested in Chapter 5, Soft Greens like Hardin view the whole of natural creation in much the same light as the zebu cow. Man must revere nature because it is in *his own* interest to do so, because it is essential to his own survival. Nature supplies us with food and medicine; it maintains breathable air; it absorbs and detoxifies our wastes. To destroy nature is, ultimately, to destroy man himself. Protecting nature is the only way to save humanity from Malthusian hell. Misanthropy in the pursuit of human happiness is no vice.

The Hard Green, by contrast, does not foresee the coming of Malthusian hell. Cows may be utterly essential to human survival in rural India, but nature as a whole is not essential to all of humanity's. We have no scientific reason at all to believe that "the balance of nature" is generally good for us; indeed, we have no reason to believe in ecological "balance" at all. The whole notion of ecological "balance" is anti-Darwinian. Evolution does not progress toward balance; evolution is the flight from it, the consequence of imbalance. There is no reason to believe that at this precise point in evolutionary history, man happily finds himself in the best of all possible ecological worlds. Nature lacks an attitude. The worst the Hard Green sees coming is Faustian hell, beige hell. The destruction of nature is an aesthetic disaster, but not a utilitarian one. Modern man can, in all likelihood, go it alone.

The green ethics you espouse are determined largely by which of these two visions you believe. Your ethics depend on what kind of future you discern through your ecological telescope and how much you trust what you see.

If you believe your Malthusian facts strongly enough, if our Malthusian vision projects out far enough in time and space, the rest is easy. It is ethical to put nature on a par with man, because man depends so completely on nature for his own survival. If you are certain of that fact, as Hardin is, you are ethically permitted—compelled even—to take drastic action. At the very least, you will avoid doing things today that will only increase suffering tomorrow. Hardin isn't obliged to feed the hungry; in fact it is unethical for him to do so, at least in places like Africa where one generation's hunger (says Hardin) just breeds another's. The numbers in the spreadsheet of future human happiness are all perfectly clear. Subtract one unit of misery today, and you add two a generation hence. Misery, like population, will only increase geometrically.

In 1974, one year into the Arab oil embargo, the facts were quite clear to all. So clear that serious American thought could turn toward a metaphor familiar to the Saudis themselves, the chopping off of hands. To some, the facts and their ethical implications are no less clear today. "Everything should be done to save the human race," Linkola insists. "Green police," unencumbered by the "syrup of ethics," must keep progress in check. He means old-fashioned ethics, of course, the ethics grounded on yesterday's misapprehension of the ecological future.

THE ETHICAL TAPEWORM

A central problem for the Malthusian ethicist is the one that utilitarians always have to grapple with: It is hard to apply their ethical prescriptions close to home. Linkola aside, no Soft Green will dare to stand up and insist on preservation of *Taenia saginata*, the "beef" tapeworm, for which man's intestines are the definitive host. Even Linkola might not care to "preserve" a particular worm that had chosen for its habitat Linkola's own guts. In the short term, and close to home, high-tech hell is usually quite cozy, in fact. Prescriptions thought up in a lifeboat really aren't. However good they may arguably be for humanity as a whole, things that feed on you are bad for you, and it is almost impossible to accept the ethical virtues of being fed upon *personally*. Judging from how they actually live, all but a very few Soft Green ethicists perceive *that* quite clearly.

Most people still understand as much when they raise their sights to survey the rest of humanity, too. Our grandparents hunted down wolves for much the same reason we still hunt viruses, to stop them from hunting us. Man's war against polio was won by displacing a potent strain of the polio virus with a weaker one bred by Albert Sabin. Environmentally, this was the equivalent of replacing Wyoming's wolves with sheep dogs, or Kenya's zebras with Masai cattle. My father worked for many years with the World Health Organization on the global campaign to eradicate smallpox. Our parents were not afflicted by HIV, but we are. So what do the Soft Greens have to say about the many schemes to endanger that particular species? Linkola will at least give you an honest answer. Few other lifeboat ethicists will; they know it is impolitic to defend those parts of nature that all ordinary people abhor. Yet if your ecological ethics forbid you to meddle with starvation in Africa, they cannot at the same time permit you to meddle with AIDS in America.

Close up in time and space, lifeboat ethics never look ethical at all. If it seems ethical to block the repair of a dam to save an endangered snail, it is only because we may sometimes wishfully believe that no humans are really endangered; when the dam breaks and children are killed, the failure to repair it seems a lot more like a moral outrage to most ordinary people. Not plowing a firebreak to save a rat seems ethical only until the fires reach real homes, not afterward. And for all we talk of the human benefits of green, money still appears to sustain people a lot better than nature does. The rich man lives healthier and longer in polluted New York than the poor man in pristine Alaska.

The Soft Green response is always to set your ethical sights farther out in time and space. When Julian Simon issued his second challenge, Ehrlich declined to bet on the "direct" measures of human welfare like life expectancy. Ehrlich concedes that life expectancies may "temporarily (that is, between 1995 and 2005) continue to rise," although the rise won't (he maintains) be "sustainable" in the longer term. Soft Greens are always doing that, always reaching beyond ordinary binoculars for a Hubble telescope of ecology, an instrument capable of peering as far as it takes—years, decades, centuries—to see beyond rosy, short-term benefits into the long-term gloom. And they are always quite confident they can see that far.

Hard Greens do not trust far-future telescopes at all. But if they do venture to peer through one they discern (in their darker moments) a quite different hell, a hell that is merely beige. Visions of beige hell do not let you dispense with the old-fashioned "ethical syrup" in the here and now. However gloomy such visions, you are still morally obliged to feed the hungry. The dire peril isn't to people, it's to redwoods and whales. Beautiful though such things are, the ethical man does not chop hands merely because hands chop trees.

INHERENT VALUE AND THE ETHICS OF PASSIVITY

Or does he? Most Soft Greens would probably accept Vice President Gore's claim in *Earth in the Balance* that nature has "inherent" value. Moral value, that is. Gore does not mean "value" in the same sense as a grand piano, say, has "value." By "inherent" he means something more, valuable not just for our sake, but for its own. Man's obligation to nature exists independently of his obligation to his fellow man. We preserve the kangaroo rat not because it is useful to people, but because nature as a whole is, in some broad sense, on a spiritual par with man.

In a passage at the end of his book *The Examined Life*, Robert Nozick suggests that the Holocaust may have changed, forever, our philosophical perspective on man in creation. After the Holocaust, after such a revelation of man's capacity for inhumanity, it is perhaps no longer correct to believe that man's disappearance from the universe would be a unique tragedy—unique because of man's special relationship with God. A tragedy, yes, but no longer a unique one. The Soft Green leads us a step farther down that philosophical path. The disappearance of other species is tragic, too, morally tragic, ethically tragic. Al Gore doesn't get into any very fine balancing here; he doesn't openly say we might on some specific occasion decide to trade the life of a child for the life of a kangaroo rat. But that is the direction of things. What else can "inherent value" mean?

In traditional Judeo-Christian ethics, by contrast, one thing is quite clear: man and nature are not equal, not ever. David Gelernter sets out the argument in his 1996 essay, "The Immorality of Environmentalism."*

> [T]he moral universe of Judaism and Christianity centers unequivocally on man. Human beings have rights and moral duties—kindness to animals being one. Animals have neither. The duty of kindness to animals is a duty owed not to nature but to God, a morally crucial distinction. . . . In the Judeo-Christian view man is emphatically not part of nature. Human life has an entirely different value from animal life, and protecting and preserving human life is a moral duty that sweeps away all "duties" to nature whatsoever—that sweeps away the very idea of "duties to nature."

Any attempt to establish moral equivalence between man and nature "isn't virtuous at all[;] [i]t's depraved." "A rich, idle misanthrope who cares more for his dogs than the neighbor's children is a study in moral degradation. So is a rich, idle nation that forces men to defer to rats."

It is no use pretending that it never comes down to that: Sometimes it does. When hunters set off in pursuit of a cougar that had killed a child in the Rockies, one protester had the candor to say what others plainly believe: There are many other children, but there are few cougars. What would Al Gore say? Declining to address the hard case, or just prevari-

---

*All my references to Gelernter are drawn from David Gelernter, "The Immorality of Environmentalism," *City Journal*, (Autumn 1996): 14–25.

cating in your answer, means you don't really have an ethical view of "inherent value" at all. Ethical norms exist for hard cases, not easy ones. *Titanic* problems are ethically interesting because those in the water are as morally worthy as those in the lifeboats. When it's people versus people, you do, sometimes, have to let ten die to save fifty. But fifty cougars? If they have *any* "inherent" moral value—in the same sense as we say that of our children—why then, at some point their claim to space on the boat begins to be weighed against the claims of children in the water.

Perhaps it should. Perhaps the Judeo-Christian moral framework is no longer right for our techno-triumphant times. Perhaps the moment has come for new scripture and a fundamental reordering of moral priorities. Perhaps nature has a moral claim to some space in the lifeboat, too. Not because man needs nature for his own survival. Not because it is convenient for man the shepherd or farmer to bring two of every other kind along for the ride, as it was on the ark. But because the other lives have moral value in their own right.

At the very least, this new view of things makes moral life very much more complicated than it once was. The old moral scheme had no trouble with man's strictly self-centered interests in protecting cattle, harvests, and such. It comes down, basically, to not coveting your neighbor's private property, and the faster we get on with issuing certificates of private title to elephants, air, and water, and so forth, the sooner we bring them back within the ambit of the Eighth and Tenth Commandments. Equally clear—in the old scheme of things—is that we may never do the same with our fellow man: Our fellow man is not our chattel, and it is never ethical to treat him as such. We are not vaguely commanded to recognize our neighbor's "inherent value." We are instructed, in no uncertain terms, to treat him as our complete moral equal.

But it takes a pretty soft green to put the moral case for environmentalism strongly enough to elevate cougars as high as children, and few do. We have moral worth; nature does, too, not quite as much as we do, but more than none. And the only trouble with saying that is that there is no intelligible moral space halfway between "equality" and "inequality." "Inherent value" is the Soft Green's vague stab at a new morality of "separate but equal."

What practical prescriptions emerge from such philosophical mush? How exactly does the moral man proceed when he believes that with every step he may crush an ant whose life has a moral worth not exactly equal to, but in some general sense comparable to, his own? Believing just that, the Jain of India continuously sweeps the ground in front of him as he walks. He is a strict vegetarian, of course. Eating plants should give him moral pause, too, if only because legions of microscopic animals dwell upon them.

Few Soft Greens literally sweep before they step, but their overarching prescription is much the same: Tread lightly, tread less, aspire not to tread at all, which is to say, be passive, stand still. All the Soft Green parables and theories converge to promote that end. The lesson of scarcity: Don't consume. The lesson of externality: Don't emit. The lesson of complexity: Don't build up the sandpile. The whole ideology: Don't produce, don't reproduce, don't plow, don't plant, don't grow, don't cut, don't hunt, don't fish, don't travel, don't discard, don't build. "Environmental activism" is, in truth, an oxymoron. The activism is always in pursuit of inactivity. If the Soft Green urges us to use solar energy or more efficient cars, it is invariably because he reviles the alternative. Most of the other things he is ostensibly for, like windmills or solar electric power, are wholly un-economic. Action is prescribed only for purposes of reaction, and the more impractical the action the more readily it is prescribed. The sum total of the moral prescription is one of passivity: systemic passivity, passivity as a daily habit, as a pervasive philosophy of life.

In its place, passivity is a good thing to pursue. Sleep knits up the ravel'd sleeve of care, and old-fashioned conservation does much the same, ecologically speaking. The main point of traditional conservation is to leave well enough alone. Define the forest, swamp, or seashore to be protected, then make sure that humanity treads lightly on it, if at all. Doing nothing, and doing it conscientiously, is often a pretty sound government policy.

But as I have argued, passivity doesn't even protect nature, not when it is pushed willy nilly, as Soft Greens push it. However much passivity we may try to impose on them from above, the great mass of people at the bottom won't stop struggling to survive. They grab what they can, burn forest, plant cassava, bear children, and flail on, just as both Malthus and Darwin said they would. It is only as an active society grows rich that people accept and welcome passivity in the well-defined context of traditional conservation, in well-defined spaces, for discernible aesthetic reasons. Passivity as a general social and political policy doesn't make things greener, it just makes them poorer, and less green to boot. The passive techno-hermit ends up not just "living in the desert," as the Greek origins of his name ("eremites") imply, he makes his own desert, the one he creates himself, by burning, plowing, scraping, and surviving as best he can. A thousand years of medieval European history should have taught us that. So stable was medieval society, so passive were its objectives, that it had no need to trouble itself with medicine, science, ecology, conservation, or anything else. Medieval Europe was passive enough for famine and plague, but not passive enough to save its forests. When they weren't busy dying, the peasants cut them all down to the ground.

It gets worse. Combine the moral duties you owe to nature with the duties you owe to the people already in the overcrowded lifeboat, and it becomes doubly moral to let Biafrans starve. Not only can feeding them "be relied upon to make matters worse" for their progeny a generation hence, it can be relied upon to make matters worse for the rain forest today. This is really all the extra moral justification you need to get from Garrett Hardin's view of things to Pentti Linkola's, from leaving human suffering alone to hurrying it along.

Which brings us to Theodore Kaczynski, Unabomber. In the "inherent value" scheme of things, just how bad a man is he? The entirety of nature, in the fullness of its moral stature, may reply that he was not really a bad man at all. Living in a wooden shack and fathering no children, he certainly lived a greener life than Al Gore, high-consuming father of four. As for Kaczynski's bombs, they merely put into practice what Hardin softly preached.

Moral man, however, must judge the matter differently. Some have particular standing to do so. David Gelernter, for example. On June 24, 1993, Kaczynski's handiwork blinded him in one eye, tore off part of his right hand, wounded him severely in the chest, and set his office on fire. Gelernter barely survived. He still writes eloquently about a wide range of subjects, ethics among them.

The Unabomber is an author, too. *The New York Times* and the *Washington Post* first published the Unabomber's 35,000-word green manifesto at the request of the FBI, in the hope, later realized, that publication would lead to his capture. But the *Times* then published an op-ed, which insisted that the manifesto was the product of a "rational" mind whose "principal beliefs are, if hardly mainstream, entirely reasonable." Society, said that author, should "treat (the Unabomber) seriously and publish his manifesto in full." The man himself should "turn to the hard business of trying to write something persuasive enough, compelling enough, to be published without homicide threats." A *Washington Post* opinion piece by Jefferson Morley, presented as "a dialogue about America's future," opened with the words: "Listen to the debate between the Unabomber and the Yale professor." Gelernter, the professor in question, would later describe the piece as "a sort of Christian-versus-lion matchup for the amusement of Sunday readers." In Lit 129, Gelernter reports, Harvard has gone on to assign Unabomber alongside Diderot, Kant, and Rousseau.

Harvard teaches Ezra Pound, too. So it should. Virulent anti-Semite though he was, Pound was also a great poet. But Kaczynski on The Green is not Pound on *The Cantos*. Kaczynski's ramblings are utterly banal; they would never have been published by anyone, anywhere, but for his bombs. In any other moral age, Kaczynski would not be on any decent person's reading list. Moral people don't burn polemical tracts, but they do refrain from attending respectfully to the banal utterances of evil men. Morley did not refrain; he amplified. In this moral age, as James Taranto would later argue in *The Wall Street Journal*, a cogent reason for putting Kaczynski to death is to separate him completely from the *Times* and the *Post*. "This man must not be rewarded for his alleged crimes by being allowed to become a pundit."[2]

I think one can only conclude that Kaczynski's green motives were what gave him positive moral weight. After all, he risked his own life for his green convictions; they almost landed him in a gas chamber. Green beliefs held *that* strongly get respectful attention. Morley, Harvard, the *Times*, don't stand for murder; they just stand for passivity in the face of murder, when murder is committed for certifiably green reasons.

Quite a bit of murder, at that. There were sixteen bombs, over seventeen years. Twenty-three people were injured, some horribly. Three died, one of them the father of a little girl too young to remember him at all. Oh yes, murder is wrong, Kaczynski must be severely punished; Morley would certainly say that, too. Everyone still *says* that. But we don't believe it, not the way we used to. Even murder can now be green.

NOT BREAD ALONE

Whether induced by respect for the "inherent value" of nature or by fear of the inherent uncertainty of life, systematic passivity is morally evil. Or so the Hard Green declares. Declares it though well aware that some ancient and venerable religious traditions teach otherwise. But our faith is not the faith of Paul of Thebes; we are not hermits or anchorites. We do not retreat from the sins of the world. We believe in the Judeo-Christian tradition of salvation through active engagement. Engagement in the world, engagement in the environment and, when it comes to that, relentless engagement against murder.

I myself don't live up to all those ideals, of course. Truth be told, I've reached a state of secure, comfortable affluence where I'm much inclined to prefer passivity over the alternative. Most particularly in environmental matters. Most of the time I much prefer nature over people. Much more to my liking, too, would be a planet that offered me the same security with, say, one-tenth the current density of humanity. I find it easy to write off people I never see. Why not make all Kansas a reserve for bison, all Tanzania for elephants and lions, and all China for pandas? I too would like to live surrounded by an endless Serengeti, a game park as large as the planet. But I also know that I'm not prepared to give up my own comfortable corner of suburban civilization. And I believe that volunteering others for the honor is immoral. Such thoughts can be dressed up in army boots, as Linkola dresses them, or in the lighter garb of the eco-hiker, as Hardin dresses them. Either way, they are evil.

However easy it may be to recognize evil, it is much less easy in our modern circumstances to define good. When people were few and weak, it was morally easy to put man's interests unequivocally above nature's. But humanity has transformed itself beyond recognition. Not biologically or spiritually, but in the knowledge and power it has acquired. However much Soft Greens may deny the fact, we have risen above nature. Not just to the point where we can destroy nature where we choose, but to the point where we can probably survive just fine in the ecological rubble.

So here we stand, with nature at our feet. How shall we rule? After a century of both brilliant creation and abominable destruction, no casual answer will do. Ancient taboos won't answer. Malthus won't answer, not even Malthus on a chip. Science won't answer. Nature itself is morally blind, as oblivious to the welfare of humans as to the welfare of our tapeworms. But as moral, sentient beings, we need not be blind or oblivious. Humans can choose. Humans must choose. We may place nature on any moral plane we wish.

We may choose, if we wish, to be pagan worshippers of Mother Earth. Before he creates Man, God creates land and water, grass, herbs, fruit trees, the seasons, and living creatures of every kind. He commands the waters to "bring forth abundantly the moving creature that hath life, and fowl that may fly above the earth in the open firmament of heaven." He creates "great whales . . . which the waters brought forth abundantly" and "every winged fowl after his kind." And these things, too, are seen to be good, and God blesses them. They, too, are commanded to be fruitful and multiply.

Or we may choose to finish the book. All in all, the Bible isn't Soft Green. As Gelernter reminds us, "Judaism and Christianity have a radical agenda; they may not live up to it in practice, but their goals are clear. They deem every human life to be sacred. At the same time, they wipe the slate clean of nature gods, nature spirits, any and all 'duties to nature.'"

Aesthetic choices are another matter. They are not sacred duties, not divine commands, they are everywhere, and always, matters of human choice. As a father, my heart goes out to the parents of the child killed by a Rocky Mountain cougar. When I imagine my own children as possible victims, I am enraged by any talk in defense of the cat. Yet I am equally sure my own children's lives will be richer for sharing this vast continent with such a beautiful carnivore. On balance, I see little in the way of redemption coming from a single cat's destruction, even after it has killed some other father's child. That is an aesthetic judgment, and a practical one, no more. I am likewise entitled—obliged—to work against the advent of high-tech hell. But for aesthetic reasons, not moral ones.

Aesthetic standards can and must be maintained, even in the context of a world in which humans still suffer. It is not immoral to build a cathedral or to preserve art in a museum, even in times when some are still homeless; one just has to think very hard about what balance to strike, and the moral imperative is still to put people first whenever direct choices have to be made. Children before cougars, when it comes to that, clearly and squarely, as it sometimes does. I can't duck my choices by saying that in the long run, the extinction of cougar young will extinguish human children, too. It's not true.

Much of the time, happily, making green choices on aesthetic grounds is a lot easier than that. Whales and redwoods are magnificent. Smallpox and tapeworms are vile. And even at that, science and aesthetic sensibility are often not so far apart. In 1977, after decades of a biochemical blitzkrieg, the World Health Organization finally won the war against the smallpox virus. Now we debate the ethics and wisdom of destroying the last few, carefully refrigerated colonies of these dreadful creatures. Debate we should. Not because smallpox is holy, but because we know that the DNA of the smallpox evolved over a very long time, that it contains an interesting, possibly important, Darwinian history. And because there should be some sense of humility, respect, admiration, even reverence for this piece of God's creation that subdued humans for three millennia before we finally subdued it in turn. We are morally entitled to make such choices precisely because our morality declares that humans are entitled to judge, to subdue, to have dominion over the Earth.

And in pursuing our aesthetic ideals, we may remember too that God's commandment to Noah was to replenish first, to replenish an Earth left empty by a catastrophic flood, and only then to subdue. There will be no more divine floods, God promises, and that covenant is not only with Noah but "with every living creature that [is] with you, of the fowl, of the cattle, and of every beast of the earth with you; from all that go out of the ark, to every beast of the earth." The power of the flood now lies in the cusp of human hands. In deciding just how much of the Earth we care to flood, we may perhaps remind ourselves that when invited to destroy another part of creation, one who came after Noah replied: "Man shall not live by bread alone."

Hardin cites that passage, too. For Hardin "the scriptural statement has a rich meaning even in the material realm. Every human being born constitutes a draft on all aspects of the environment—food, air, water, unspoiled scenery, occasional and optional solitude, beaches, contact with wild animals, fishing, hunting—the list is long and incompletely known." And off he runs. For Hardin, one of the holiest phrases in the Bible inspires only another neo-Malthusian rant.

Perhaps Hardin believes in nothing more. Perhaps he shouldn't. Science, after all, has shouldered aside much of traditional religion. From Copernicus to Darwin, as Carl Sagan noted, the advance of science has been a "series of Great Demotions, downlifting experiences, demonstrations of our apparent insignificance." Darwin himself understood as much, even though he ended *The Origin of Species* trying to discern a new grandeur in what his book revealed. In 1953, *The Double Helix* reduced us to fungible molecules: beautiful ones, of course, quite as beautiful as a toad's. As others have pointed out, Darwin was whistling past the graveyard. Kant's philosophy put grandeur into life: human life. Darwin's science took it out. Small wonder that Darwin was not welcomed by the established Christian church. Established religion has been in decline ever since.

New faiths have risen to occupy the space left vacant. Communism and fascism had their vicious days. Now we have the green reaction, a new creation science, but without a Creator. "One of the keys to understanding the twentieth century is to identify the beneficiaries of the decline in formal religion," historian Paul Johnson observed in a prescient 1980 essay.[3] "The religious impulse—with all the excesses of zealotry and intolerance it can produce—remains powerful, but expresses itself in secular substitutes." A new creed has evolved to provide an "emotional outlet for educated, middle-class opinion." Johnson's 1980 focus was on the west's rapidly evolving preoccupation with health and safety regulation. Since then, environmentalism has synthesized those impulses into a general, new-age philosophy of good and evil. The good centers on what Johnson described as a "quasi-mystical vision of total purity." The evil, on whatever creations of man are thought to corrupt it. The new creed "combines fear of technology, hatred of capitalism, and a compulsive itch to interfere." It is obsessed with whatever current subject unites the "maximum of public apprehension" with the "minimum of public understanding."

At its most reductionist, Soft Green reverts back to sacred-cow taboos. It engages the sociological tools of the ancient Bushman to resist the technological tools of the modern capitalist. These ancient sociological tools were once useful, when experience moved far out ahead of understanding. The Soft Green reverses that intellectual process: He claims, through his future-gazing computers, to understand what cannot be measured or experienced. We are to shun the technological pig, the pesticide, the nuclear fuel, not because it made us sick yesterday, but because the model says it will make us sick two decades hence.

The objects of good keep changing. Birds, whales, rats, frogs, snails, redwoods, and desert pines. The targets of evil, too: nuclear power, dioxin, plastics, diapers, CFCs, and carbon dioxide. Some endure, others are much in debate for a month or two, then fade from sight. In 1980, as Johnson pointed out, nuclear power stood as "the new Sin against the Holy Ghost—radiating evil, as it were, over the whole planet and, like Original Sin, even infecting future generations." Alongside it, the chemical carcinogen had become "the universal, ubiquitous, omnipresent spirit of Satan, threatening to poison all with its corruption." The new taboos are all dressed up as science, all polished and buffed by computer, the most powerful construct of modern technology. But they are, in fact, a passionate reaction against science's complete triumph over nature.

Like the other two secular religions of this century, Soft Green is preoccupied with ineluctable historical imperatives, the grand sweep of things, not day-to-day or even year-to-year, but decade-to-decade, century-to-century. Seizing intellectual control of the long-term factual future is politically powerful. But it is the opiate of the evil politician, an ethical poison that can end up justifying any amount of viciousness in the present, for the good of the distant future. Stalin and Hitler had their sweeping future-history facts, and even a vaguely proto-green vocabulary ("racial purity," "lebensraum") to go with them. Big, sweeping facts, the whole expanse of history unfolding—not in the past, but in the future. They knew what the future held, and to whom it belonged. The future belonged to them.

Soft Green is a green disaster and an ethical disaster, too, because its facts are too numerous, too weak, too thin, too long term, too easily reinvented, too malleable, too changeable. Hardin's world is morally corrupt because Hardin gets to make up his future facts as he goes and project them as far out in time as he likes, out where nobody can grapple with them seriously or prove them wrong. The lifeboat is twice as crowded but still well afloat a quarter-century after he wrote. It makes no difference, of course: Hardin can simply tell us to wait a quarter-century more. Hardin is so very certain about his future that he knows it is wrong to feed the hungry. He can even cite scripture to explain why.

The Limits
of Certitude

◆

"Think globally, act locally," the Softs once advised us. "Small is beautiful," they added. But what they really meant was think through the big computer model, act through the big regulatory agency, enlist the federal government, co-opt the United Nations, convene in Kyoto, prescribe from the top down, think globally and act globally, too. Perhaps aphorisms are too easily misunderstood to be fair game in green discourse. But if there is to be a Hard Green alternative, it must be: Think locally, act locally, and we really mean it. Our solution for scarcity: markets. For pollution: property. For complexity: evolution. For efficiency: markets. For wilderness: wilderness itself. Small answers. Small models. Small procedures. Small government. The hurried reader who has skipped directly to this chapter need not in fact plow through all the others. They all come down to this. Small is indeed beautiful. Softs just don't mean it. Hards do.

Softs are the mega-thinkers of our day, the ones who grasp it all and are prepared to prescribe accordingly. The world has seen their kind before. Natural man is inherently good, declares the French philosopher Jean-Jacques Rousseau in 1762. In 1793, William Godwin foresees a millennium in which rational men will live prosperously and harmoniously without laws and institutions. A year later the Marquis de Condorcet publishes a theory of continuous human progress. Having advanced through nine great epochs, mankind will attain ultimate perfection in the tenth. Karl Marx sees things much the same way in 1848. The capitalist "prehistory" of human society, rife with contradiction and alienation, will soon reach its "general crisis," then collapse like a sandpile. After that, abundance. The theory of scarcity propounded by that "miserable parson," Thomas Malthus, is a "repulsive blasphemy against man and nature," Marx declares. Better politics will solve everything. "The golden age is coming," Lenin proclaims when the Bolsheviks seize power in 1918. "People will live without laws or punishment, doing of their own free will what is good and just."

Big-futurists are still among us, and not all of them are Soft. Many now see technology as the prime mover of the future. "We have entered a period of sustained growth that could eventually double the world's economy every dozen years and bring increasing prosperity for—quite literally—billions of people on the planet." So declare Peter Schwartz and Peter Leyden in "The Long Boom,"[1] a 1997 article in *Wired* magazine. Four "great waves of technology" propel huge increases in productivity and extraordinary growth; a fifth saves the environment. Chips and Internet experience "explosive," "exponential" growth. Biotechnology transforms medicine and agriculture. The new technologies operate at low temperatures, emulating nature and creating much less pollution. Information technology saves energy and resources, too. Then comes the "fifth wave," hydrogen as an alternative source of energy.

All megatrend futurists sound much the same. Sweeping vision, far future, exponential change, grand prescription; epochs, waves, general crisis. The human mind can assign no fixed limits to its own advancement in knowledge and virtue or even to the prolongation of bodily life. So says the Marquis de Condorcet in 1794; so says *Wired* magazine today. Or flip all the highs to lows, and you have Thomas Malthus or Paul Ehrlich. Technophilic libertarian optimist and technophobic socialist pessimist are equally certain of one thing: The Future Belongs to Us.

They know, because they think scientifically. Rousseau and Godwin insist they are thinking "scientifically." Marx proclaims an integrated, "scientific" doctrine that applies universally to all nature. They understand history and technology. Their successors understand ecology and computers. They grasp the world dynamics of scarcity, the statistical dynamics of disease, the temporal dynamics of disaster, the complexity dynamics of the sandpile. They grasp the long term.

Like his contemporary, Rousseau, and like Marx a century later, Adam Smith is equally determined to explain the grand sweep of history. He concludes that civilization progresses through four main stages: hunter, nomadic agriculture, manorial farming, and finally the commercial interdependence of nation states. Institutions and laws evolve apace: private property, feudalism, and finally laissez-faire capitalism, the system of "perfect liberty." The parallels with Marx's view of history are remarkable. But the parallels end when it comes to the future. Marx discerns everything that lies ahead; Smith discerns nothing.

For Adam Smith, the prime mover of history is the individual, the basic, personal determination to improve one's own lot. The individual is not interested in the collapse of sandpiles in the next century; he is interested in tomorrow's lunch. Rousseau, Condorcet, and Marx claim to discern the great currents of history. Adam Smith begins *The Wealth of Nations* with a long discussion of the efficient manufacture of pins. Rousseau and Marx present us with the image of Man in Chains. Adam Smith's most quoted passage is about what impels butcher, brewer, and baker to supply us with dinner.* The modern Marxist still sees capitalism as systematic, orchestrated theft. The modern economist maintains that there is no system to stock market prices at all, only a "random walk" and that the "efficient market" knows far more than any individual actor.

The other great slayer of futurism is Charles Darwin. He too is a genius of the small—he ends his professional life, as he begins it, writing about earthworms. Marx announces a new science of all nature; his contemporary, Darwin, ventures one, too. But Darwin's is as modest as Adam Smith's, a description of historical process, not of future result. Biological history has no end in Darwin's books; it doesn't even have a direction, it is a random walk through an indifferent environment. Watson and Crick describe the rather simple chemistry behind this random walk of history a century later. Tiny genetic changes are found to account for a great deal of variation: mutate 1 percent of the genome to transform an ape into a human. Marxists insist they much prefer the evolutionary science of Lysenko.

One school of the future is confident and self-assertive. It knows what lies far ahead and just how to shape that distant shore. It has all the information it needs to make the future a better place; the one thing more

---

*"It is not from the benevolence of the butcher, the brewer, or the baker that we expect our dinner, but from their regard to their own interest. We address ourselves, not to their humanity but to their self-love, and never talk to them of our own necessities but of their advantages."

it needs is political power. The other school is diffident. Its theory establishes only that the trajectory of the far future is unknowable and therefore beyond direct control; it is a theory, really, about the limits of theory. So it prescribes means, not ends. It is confident only that better things emerge from better process. And that better process is small and decentralized: It centers on free markets, not on prescription.

SCARCITY

Marx saw looming scarcity, too. What the capitalists were going to diminish and ultimately exhaust was employment for the common man. Capitalists would invest more and more in their machines, making them ever more productive. The productivity of capital would rise relentlessly, and the ranks of the unemployed would rise apace,* eventually to socially destabilizing levels. The unemployed masses would rise up and seize control of the machines.

The three great green theories of collapse are all variations on the same story. It's not jobs that capitalism will exhaust, it's the lumber and coal on which it feeds, say limits theorists like Malthus and Forrester. Or the air and water, say the Nuisance Cassandras: Capitalism will eventually choke on its own wastes. Or capitalism will build up its brittle technology to unstable heights and collapse will inevitably follow once again. Capitalism, it seems, is always approaching its final crisis, because it is too efficient, too voracious, too befouling, too technologically bold, or, in the most recent collapse theory, too inefficient.

The dress is looser, the computer-applied makeup is thicker, but the strumpet underneath is the same old Malthus, still walking the streets of public policy. All the Malthusians have learned is to cloud their auguries in imprecision. They are very certain about what they know, but they make few quantitative predictions. They know calamity is coming, and sooner than the rest of us may suppose, but they offer no clear timetables, at least not short ones. They used to talk decades, but now prefer half centuries, or longer. Such timetables certainly minimize the personal price they pay for

---

*Soft Green neo-Marxists still believe this. See Jeremy Rifkin, *The End of Work: The Decline of the Global Labor Force and the Dawn of the Post-Market Era* (New York: Putnam, 1995). Rifkin argues that silicon-powered machines will eliminate most of our jobs. The machines will leave behind teeming masses of the discontented unemployed. Only a tiny nucleus of the techno-elite will prosper and rule. Rifkin, a passionate opponent of biotechnology, is a sandpile theorist too. See Jeremy Rifkin, *Algeny* (New York: Viking, 1983).

being wrong. Who today takes the blame for silly, next-century predictions made by some ancestral fool who voted for William Jennings Bryan in the election of 1900, when Victoria still reigned?

So far as market scarcity is concerned, Adam Smith's description of the future is the only one that has endured, because it describes only a durable process, no more. Channel the acquisitive drive of the individual through the laws of the free market, subdivide labor, specialize, develop new skills, and national wealth will grow. Adam Smith, scholar of the preindustrial English pin factory, sees no limit to that process, and none has been found in the two centuries since. Substitute "chip" for "pin" and Smith describes today's global information economy in terms quite suitable for publication in *Wired*. More suitable, indeed, than long-boom articles predicting geometric growth of all things bright and beautiful.

The futurists are always proclaiming the end of an epoch, the death of some old order or other. But it is the epoch of the market-scarcity futurists themselves that should finally draw to a close. It has had a good run, from Malthus in 1798, to *The Limits* in 1972, to the Simon-Ehrlich payoff in 1990. It is time, now, to stop building big models of the future of the market. Free markets are beyond modeling. Experience merely teaches us that they don't run out of anything that people are able to define, contain, package, and trade.

Experience teaches the opposite about wilderness. Absent conscious, deliberate husbandry of some kind, the swamp is drained, the sea is fished clean, and the wilderness is leveled and paved. There is only one great green scarcity to address, the one T. R. already discerned a century ago. No big computer model of the future is required to perceive this scarcity, and no vast government bureaucracy is required to deal with it. Traditional conservation is transparent in its logic and simple in its methods. We conserve the wilderness because it is beautiful. We conserve it today, with today's means, for today's ends, and merely hope that posterity will be grateful that we did.

EXTERNALITY

Much natural wealth dwells, however, outside the direct protection of the Invisible Hand. This creates problems that are altogether visible and real. The only debate is about how best to solve them.

People who know the future and how best to manage it see private initiative and control as the problem, not the solution. Humanity was happy when the Earth itself belonged to all; private property is the "destructive" and "fatal" invention responsible for all subsequent "horrors" (Rousseau). Private property is "the tangible material expression of alienated human life"; capitalist production depends on "exhausting at the same time the two sources from which all wealth springs: the earth and the worker" (Marx). Even the new techno-libertarians seem to view property as obstructive and futile, too, at least in the new realm of cyberspace. "Open, good. Closed, bad. Tattoo it on your forehead," says *Wired*. Whatever that may mean, it doesn't sound like "No Trespassing."

Hard Greens see dispersion as the problem, not property. A very difficult and pervasive problem, too. Virtually every advance of technology, and every major law and institution of capitalism, is directed at resisting drift, dispersion, disorder, chaos: externality, broadly defined. The whole point of private property is to resist theft, the oldest economic externality of all. Adam Smith understands quite as well as Rousseau or Marx that private property is instituted for the defense of "those who have some property against those who have none at all." Property is not an instrument of social equality, but it assuredly is an instrument of containment. Containing pollution fits quite naturally with the instinct to contain, protect, and preserve.

But the Hard Green has to proceed more cautiously than the Soft, not because he discerns fewer externalities, but because he discerns so many more. Not just smokestack, sewer pipe, and synthetic chemical, but nature's biocides and dispersants, too. Dispersion is the ineluctable imperative of life. Hyper-successful species that they are, capitalists pollute, despoil, and dump a great quantity of effluents on their surroundings. Other species can't begin to match us, but they do the best they can. The Hard Green accepts that Darwinian truth, too. There may be a cure for AIDS somewhere up there in the trees of the rain forest, but trees are also where the HIV virus came down from in the first place, by way of monkeys. The totality of nature has not been mystically perfected for human benefit. Darwin teaches otherwise.

Command and control is what you use when you have in hand a seemingly precise model that tells you everything you need to know about pollution, its consequences, and its control. Markets are what you turn to when you don't. Markets require social structures, but they do not require elaborate models. The less we know, the more we need them. And in the overall battle against dispersion and decay we are very ignorant indeed. The problems are ubiquitous and refractory. The costs of control vary enormously. The possibilities for innovation and improvement are endlessly diverse.

Assigning private title is the most neutral, general, and effective way to promote containment. Defining new property rights requires the least amount of new information and none at all about the far future. Defining property rights is itself a process, and it can be a politically neutral one. Most of the private property we own today was invented in the common-law courts of England. The job isn't finished, of course; we will have to continue inventing new forms of property as fast as we discover new forms of value, whether in the digital recesses of cyberspace or in the green expanses of forest, lake, and ocean. Information technology offers us ever-improving ways to meter and track bad things as well as we track goods, new power to give a lot of pollution what it has always lacked: a face, a name, an owner. Once you can do that accurately, the rest is legal detail.

Or it will be, if we hold the line against the pernicious forces of political dispersion, against the people with all the facts in hand who are always grabbing for public power to go with them. Minimizing externalities requires information, coordination, wisdom, and knowledge, on a scale that central planners can never begin to acquire, however many computers they may have in hand. It is precisely because externalities are so ubiquitous, and information so imperfect, that we insist on market solutions. No other approach will end up solving anything.

COMPLEXITY

The Soft Green professes ignorance about the future in only one arena, and even there, true to form, he professes too much. He is certain that he doesn't understand every last perverse possibility of high technology and that no one else does either. Nuclear power is inherently unstable; the side effects of pesticides and genetic engineering can never be contained or predicted either. The Soft Green can predict the future of all the resources in the world, but not the future of a solitary nuke in Tennessee. Backed by the full power of the most advanced computers, the Soft Green confidently declares that high technology cannot be trusted. Yet even here he knows more than anyone else. He knows sandpile technology when he sees it. He knows that sandpiles always collapse badly, never well. Except for nature's sandpile, which is stacked in humanity's favor.

This century's Marxists were equally certain about the future of technology, except that they were fanatical techno-utopians; for them all the downs were going to be ups. There were no limits to industrial progress, no social ills that could not be cured with a new dam, a new power plant, or a new salmon canning factory in Irkutsk. More coal, more steel, more engines, more smoke stacks, would lift humanity up and up. And so they did, in a small minority of countries governed by what Hard Greens still call civilization. In the rest, technological progress only amplified the power of the gulag and the gas chamber.

Put aside the fact that they have stepped through a mirror, and Soft Greens today do not look very different from Soft Reds of yesteryear. The new Softs are every bit as certain about the techno-sandpile as the old, the only difference being that the old ones said it would inevitably rise, while the new say it must inevitably fall. The old embraced the technology that was bound to make goods like iron and electricity too cheap to meter. The new embrace only technology like solar power and windmills that is too expensive to matter.

Hard Greens will have none of either, not the high-tech utopian nor his dystopian twin. Most things aren't sandpiles; those that are reveal their propensities in only one way, and that is by collapsing. The past is the only reliable predictor of the future here. In democratically governed nations with free markets, the nukes have not collapsed willy nilly; the technological past has been quite stable.

Nature and the planet, too, seem to have found their own ways past collapse, not entirely gentle or even, but altogether durable nonetheless. Al Gore notwithstanding, there seems to be little of the sandpile—no inherent, fragile, instability—in our biological and geophysical surroundings. Lovelock may be quite right: "Our uncertainties about the future of our planet and the consequences of pollution stem largely from our ignorance of planetary control systems." They may be a lot better than we suppose. It may be mostly honey out there, not sand.

Techno-complexity certainly isn't green in and of itself. But the technologies of the third dimension, the ones that move us away from the interface to find energy and space far from the surface where the life dwells: Those technologies are all complex. Complexity isn't sufficient to keep us green, but it is certainly necessary. It lets us grow more food on less farmland and extract more energy from less surface. It lets us move people in the lifeless stratosphere, instead of laying iron tracks across the prairies. This conclusion is solidly grounded on direct observation and experience. It does not take a big computer model to prove that high-tech agriculture consumes less land than low-tech does, or that extracting energy from deep in the Earth is kinder to life at the interface than are surface-devouring alternatives.

This is not to say that we must embrace complexity wherever we find it; far from it. Some complex technologies are green but others aren't, and in any event, green objectives are not the only ones we wish to promote. People are generally happier with familiar things that they feel they understand and control. James Thurber's grandmother "lived the latter years of her life in the horrible suspicion that electricity was dripping invisibly all over the house," her grandson wrote in a 1933 essay. She spent her days turning off the wall switches that lead to empty sockets, "happy in the satisfaction that she had stopped not only a costly but a dangerous leakage." She was undoubtedly much happier with her oil lamps and wood stove. She was entitled to be. She just wasn't entitled to argue they were better for the environment, or safer for her own household, not if in fact they weren't. To the extent one dares to generalize about such things at all, it often seems to take more complexity to make things safer, cleaner, better in every other dimension except the psychological. Our reflexive modern hostility to high technology has been, by and large, an environmental disaster.

And we must remember, too, that simplicity in the things we create ourselves is itself beautiful to many eyes. A Shaker chair and a Norman Rockwell illustration are more beautiful, to my eyes at least, than Victorian lumber or abstract paint splatter. There is something good about simplicity. Not safe, not stable, not efficient, merely good. It is aesthetically pleasing. In engineering it is elegant. In science it is beautiful. Occam's razor, the underpinning of all real science, exhorts us to discover the simplest possible explanations for complex phenomena. Simple is better, at least when it eliminates clutter. Thoreau was right, not about science or system engineering, but about life. Complexity comes too easily and demands too much. Simplify. It is a gift to be simple.

The opposite can be said of the life that surrounds us, of the things that merely are. Complexity in nature is beautiful. Not safe, not stable, but beautiful because . . . because we find it so, and for no other reason. The toadstool and the venomous snake, too, alongside the passion fruit. There may be nothing more out there than a great deal of fascinating beauty in complexity itself.

With nature, stability is beside the point. Slashing and burning the rain forest is ugly. It is aesthetically abominable. We should revere life on Earth not because we fear catastrophic failure but because life is a good that requires no further justification. Do we really need more? Cu Chi today is immeasurably more pleasant and beautiful than when we left it two decades ago. The genetic code of life on Earth is a gigantic biological manuscript that we have scarcely yet begun to examine or decode. It is a record as ancient as life on Earth, encrypted in molecules scattered across the vast, delicate, shining, abundance of the canopies, trunks, stems, and soil of the rain forest. It would be a great pity—an aesthetic pity—to burn such a book before we had even found the time to read it.

Efficiency

Marx himself allowed that capitalists were efficient. Efficiency was what was going to limit capitalism's own march through the pages of history. More capital meant ever-more efficient machines, which meant ever-increasing unemployment. The Soft Green tells exactly the same story today, but likes it: More efficient refrigerators mean less employment at the power plant and the mine-head. Now as then, the logic seems impeccable—on paper. Yet the real world does not see it. Efficiency rises and rises, but so do employment, consumption, power plants, and drilling rigs. The scarcity futurists somehow miss something so fundamental that what is supposed to go down goes up.

Seeing what lies ahead, seeing it so clearly and at such a great distance, megatrend futurists always have a lot to say about how best to make the journey "efficient." They understand clearly how all the economic currents interact. They understand how this trickles into that, how fiddling with this economic spigot over here will produce that excellent consequence over there. Much as they reviled efficiency under the capitalist's private control, the Marxists were boundlessly confident in their own ability to deliver it, through "scientific" central planning. Stalin "seemed to live in a half-real and half-dreamy world of statistical figures and indices, of industrial orders and instructions, a world in which no target and no objective seemed to be beyond his and the party's grasp."[2] The party placed grand orders, for steel, electricity, and wheat and for environmental protection, too. Everything was meticulously modeled and calculated and planned, down to the last detail. What actually emerged was lunacy, both economic and environmental. The political planners of the scientific future delivered a random walk to waste, inefficiency, dispersion, and disaster.

Happily for the environment the Softs lack Stalin's totalitarian grip on things, but they still go for as much as they can, with what grip they have.

And unhappily for the environment, they have quite a bit. They too have their reams of efficiency statistics, which they translate into prescriptions for the gas economies of cars, the electrical economies of refrigerators, and the aquatic economies of toilet tanks. They understand how the efficiency they prescribe for the refrigerator will trickle back up through the power plant to the mine-head; they understand the intricate trickling in ways the ordinary householder or factory manager will never grasp. They alone understand why protecting the wilderness is not enough; to protect the wilderness they must redesign the refrigerator, too. They don't just move Yellowstone into the public domain, they move the toilet tank industry there, too.

This means less free market, not more, and therefore—quite predictably and inevitably—less efficiency, not more. Which means less wealth, not more. Which means, in the end, less green, not more.

Wealth is green, poverty isn't. Once again, that conclusion does not emerge from elaborate models of the future, it is compelled by direct observation of the past and present. Poverty curbs environmental decay the Malthusian way; wealth curbs it, too, the way T. R. prescribed. Wealth solves the problem of scarcity with abundance. Wealth solves the problem of population by defending life, not by surrendering to death.

And wealth eventually pours into green, for the simple reason that it has nowhere else to pour. As I wrote in Chapter 9, the only form of poverty that is green is the poverty chosen freely by the man so rich—so rich in capital, so rich in spirit—that he freely chooses to spend his treasure preserving and advancing more life than his own. Green is what people become when they are secure, when their normal appetites are satiated, when they are confident of the future, their own and their children's. It is wealth itself that gives ordinary men the confidence to be generous to the world beyond.

This is why Hard Greens must resolve in the end to be inefficient. Inefficient in government. Inefficient in prescribing and controlling and dictating, as Stalin did, as Softs still aspire to do. Wealth is not the product of efficient central planners, it is the product of unplanned markets. Markets, we well know, are appallingly inefficient. How could they be otherwise? In free markets, competitors duplicate each other's investments, businesses rise and fall, many go bankrupt, and everyone's information about what's going on is miserably incomplete. Private projections of the future are laughably ad hoc and inadequate; they do not look the least bit professional or serious when compared, say, with the enormous modeling and computing power that produced *The Limits to Growth*. No, free markets are not efficient at all. They are just less inefficient than every other alternative. So they generate wealth much, much faster. And wealth is green.

ANARCHY AND UTOPIA

"Man was born free, but he is everywhere in chains," declares Rousseau in the opening sentence of *The Social Contract*. Liberty is to be found in obedience to a self-imposed law. Society is to be united by a "general will," however it may conflict with personal interest. But most people are stupid, so they need a lawgiver, one capable of "forcing a man to be free." The Calvinist republic of Geneva—where Rousseau has taken political refuge, with his laundry-maid mistress—represents the best balance one can strike between individual autonomy and civic life. Godwin, half anarchist, half communist, arrives at much the same conclusion. Every form of "co-operation . . . is in some sense an evil," but it is also inescapable. Godwin, too, ends up favoring small self-subsisting communities, with property owned in common.

People who trust models more than markets always end up wrestling with how to reconcile freedom with cooperation. They know, as Adam Smith knew, that man is instinctively selfish; they also know that he must cooperate, but they wonder how to channel cooperation toward the right ends. The right ends aren't the market's, they are the ends defined by the philosophical mind, or, more recently, by the philosophical computer model. Those ends have to be imposed on the unphilosophical rest, digitally clueless knaves, fools, and vandals that they are.

How much prescription you are willing to impose thus depends on the limits, if any, of your trust in government. Adam Smith trusted markets more; indeed, he believed that the main cause of market failure was government being too willing to enlist with favored private interests in suppressing competition. Rousseau and Godwin at least recognized some limits to the beneficence of government, too. Marx and Lenin didn't. When you are very certain about what will be and what ought to be, you are willing to prescribe a great deal.

It is that frame of mind, too confident about its vision, too prescriptive in its politics, that transforms worship into witch hunt, godliness into Inquisition, and the rule of law into the rule of bureaucratic labyrinth. The Softs don't oppose traditional conservation, of course. Quite the contrary, they will insist: They want to push it much farther. But then, that is what Rousseau said about liberty, Marx about wealth, and Lenin about democracy. As zealots always do, they all promised more of the same: much, much more. And they offered an incontrovertible political prescription for delivering. But with liberty, wealth, democracy, and conservation, too, incontrovertible prescriptions only destroy. You cannot force man to be free, as Rousseau proposed to do. Nor can you hector, regulate, and impoverish him into respecting life around him.

Traditional conservation requires no totalitarian ideology of green. It does not force itself into the ordinary economic sphere of life. To the contrary, it helps affirm and delineate the basic division between private space and public. It is just Central Park, but on a much larger and more varied scale, public spaces set aside for common enjoyment and noncommercial use. And on this vast continent of ours, with its richly varied wildlife, there is much worth conserving for those ends. Old-fashioned conservation fits comfortably with the Hard Green's political prescription for ordinary pollution: Convert big pollutants into private property and ignore little ones. Conservation requires government to set aside and leave alone delineated segments of land, forest, ocean, and wildlife. It does not invite government to micro-manage the myriad details of human life. It is a leave-things-alone philosophy for certain delineated spaces, not a philosophy of systematic submission for the populace as a whole.

IDOLATRY

With such a secure grip on the hereafter, the futurists have no use for holy writ or "miserable clerics." Christianity is "useless" (Rousseau, Godwin, Condorcet), "an opium for the people" (Marx). But they recognize that mobilizing the masses requires a psychologically compelling creed of some sort. Rousseau suggests a new secular religion. Marx delivers. Marx publishes *The Communist Manifesto* in 1848, the same year that Charles Dickens publishes *Dombey and Son*. Both are concerned about wealth and poverty; both writers feel for the poor. But the author of *Oliver Twist* (1838) and *Hard Times* (1854) sees individual suffering in front of his own eyes, and with the power of his pen, changes it for the better. Marx is all statistics and theory, and what emerges from his powerful pen is a manifesto for evil. Dickens sees Scrooge and Tiny Tim, and writes *A Christmas Carol*. Marx sees nothing but opiate of the people. Dickens sees suffering, and advocates charity. Marx sees suffering and advocates blood.

When the distant future is so clear and certain, and so very much under the control of your very own eyes, you can dispense with God, and your ethics can become very accommodating. Master of the lifeboat and clairvoyant of the high seas, you do what has to be done to ensure the greatest good for the greatest number. Knowing how the future will unfold, you are able to keep the sorry details of the present in perspective. You balance the suffering that you tolerate (or even inflict) today against the larger quantum of suffering you will avert in the next century when some computer model has finally run its course.

Knowing far less about the future leaves less leeway for moral maneuvering in the present. If you clearly understand the most ephemeral causes of the most difficult diseases, you confidently advise people all about them. People less clear in their own minds about all the mysteries of nature consider it immoral even to suggest that significant, long-term hazards to health lie in things like trace synthetic chemicals or polluted air, far beyond any individual's direct control. Immoral because such stories may encourage people to neglect the many serious and unambiguous hazards they can control, like tobacco, sex, seatbelts, and bad diet.

It is much easier to persecute people—it's not really "persecution" at all—when you simply *know* they are poisoning the wells. It is, likewise, quite easy to justify almost anything through very elaborate models that tell you exactly how today's choices will affect the Index of Total Human Happiness a century from now. Garrett Hardin and Pentti Linkola are perfectly ethical, given the one future they see so clearly and the many alternative futures they decline to see at all. Hard Greens can respond only that it is immoral to see such futures because they are not facts; they are just doleful speculation, evil programmed with all solemnity into the spreadsheet of a computer.

UNCERTAINTY

The Soft Green is quite certain about the ecological destiny of the planet: the "carrying capacity" of the Earth, the "exhaustion of resources," the poisoning of the biosphere, the extinction of species, the change of climate. The Soft Green, who trusts models, claims to know what the Hard Green, who trusts markets, believes to be unknowable. But the more agnostic the Hard Green's views about the far future, the more zealous the Soft Green's faith in his prescriptions. His model tells the Soft what is surely coming, and it stands uncontradicted by anything stronger than doubt. The humbler the origins of the Hard Green's faith in the market, the prouder the Soft Green's faith in his model. The divide is not between the reckless and the cautious, still less between those who would act and those who would stand by and watch. It is between those who are certain they know certain things and those who are certain they don't.

Yet for all his many, high certainties, the Soft Green gets a lot of intellectual mileage out of uncertainty, too. Highly certain though he is about the ecological destiny of the planet, the Soft Green professes boundless uncertainty about the technological future of the nuclear power plant, pesticide, and fossil fuel, as he does about the long-term effects of fluorocarbons, dioxins, and carbon dioxide. That uncertainty, he insists, is reason enough to ban them all.

The Soft Green now gives his theory of ignorance a solemn, scientific-sounding name: The Precautionary Principle. He grants that his model can't say for sure which of several scenarios of the future will unfold. But comparing the costs and benefits of each makes clear what must be done. Curtail nuclear power, because however low its likelihood, the risk of a meltdown is intolerable considering the enormity of the (hypothetical) consequences. Curtail fossil fuels, because living in a greenhouse will be very costly indeed, if by chance it comes to that. The one enterprise The Precautionary Principle never extends to is the political enterprise. Nothing to be cautious about or to curtail there.

The Hard Green orders his precautions the other way around. If you are going to trust models anywhere, trust high technology first. Compared to a national economy or a rain forest, a power plant is simple, well defined, and designed for stability. We readily stake our lives on that design process every time we board a plane. Nature's stability is the product of long evolution, a highly stabilizing process, too, but still a blind one, one that is not encased in concrete or steel. It is in the political arena that the Hard Green feels most precautionary of all, having learned how easily political solutions can make bad things worse. When Precautious Softs throttled nuclear power in the 1970s, they thought the Earth was cooling and that we would soon run out of fossil fuels. Today, we burn more fossil fuel than ever, and they say the Earth is warming. Perhaps all would have been well if they had managed to throttle coal, too, but they didn't. The Precautionists somehow failed to predict their own failure in the arena they trust the most, the political arena.

Moreover, the Soft Greens had no use for the Principle when it traveled under its original name, cost-benefit analysis. They excluded any consideration of cost from the 1970 Clean Air Act Amendments. They were outraged when Ronald Reagan issued a general "cost benefit" executive order in 1981. Limit enormous expenditure in pursuit of trivial risks? The numbers are far too soft to be of any use, the Soft Greens insist. Limit enormous risk at modest expense? Why of course we must, the Softs declare: The numbers are quite convincing.

Curiously, the Soft Green's ecological certainties often emerge from the very same machines—large computers—that Hard Greens use to design nukes or bioengineer crops. The Soft Green often reproaches the Hard for over-confidence and techno-arrogance. Yet many a Soft Green puts at least as much arrogant faith in his camp's models of the ecological future. One camp predicts the future performance of a nuclear power plant or dam, the other predicts the future performance of a global economy or atmosphere.

The Hard Green is inclined to think that uncertainty is both greater and more perilous when it lurks in the Darwinian rain forest than when it lurks in the modern engineer's concrete and steel. More generally, however, the Hard Green accepts that the future is uncertain, accepts that far more completely than the Soft. Without any precise spreadsheet of the future in hand, he hesitates to trade common decency today for the salvation of humanity fifty years hence. Uncertainty does not tilt his views one way rather than another. No law of science, no principle of ecology, declares that with five billion hungry humans already on the planet, abolishing nuclear power or bovine hormones is greener than abolishing biomass fuel or catalytic converters. Action can produce happy surprises; inaction can produce unhappy ones. Hard Greens do not believe in any great Principle of Regret. Uncertainty doesn't resolve economic, ethical, or political quandaries. Ignorance does not tell us to favor one policy over another. Ignorance is just ignorance.

Caution is a perfectly sensible attitude to cultivate, even without tarting it up in the makeup of pseudo-science. But you can't wring much useful policy out of it. The geophysical environment stores and cycles thousands of different atoms and molecules. The biological environment, millions more, including those produced by things like salmonella, which has its own views of what is "cautious." Each of the countless millions of species on Earth is defined by its own unique chemistry. Humans themselves are biochemically differentiated by sex, age, and ethnic group. Every biochemical on Earth can react well, badly, or not at all with every other, and with the eighty thousand synthetic chemicals in common commercial use. There are trillions of possible interactions. That does not mean we should simply ignore the possible toxic effects of chlorine in our drinking water. It does mean that trying to pluck policy out of the depths of a Precautionary Principle is a waste of time.

And calling it a Precautionary Principle to imply some serious science behind is an outright fraud. Big-future theories are never scientific. Karl Popper fingered Marxism and Freudian psychology as two paradigmatic examples of claims dressed up as "science" that were not in fact "falsifiable." Proponents of these views could find confirmation in everything and refutation in nothing. "A Marxist could not open a newspaper without finding on every page confirming evidence for his interpretation of history; not only in the news, but also in its presentation—which revealed the class bias of the paper—and especially of course in what the paper did not say."* To be scientific a theory must make predictions concrete enough to be proved wrong if the claim is not in fact true. Very little indeed in the intellectual arsenal of the Soft Green meets that requirement. Science breaks down when the scenarios get too long, when the search reaches too far into the depths of time and statistical mist.

For the Soft Green, however, this is often the best possible outcome. All the trappings of science remain, to keep the business solemn and public respect high. But little if any of the discipline. In the end, the Soft Green just conjures up demons wherever he likes, and gets on with exorcising them. The process is dressed up as science, but it is irreducibly political. It is politics by other means. It is a system perfectly designed to fund and grow the critical establishment, the legions of academics and bureaucrats whose occupation it is to imagine, worry, and prescribe.

When we are sufficiently selective about our grave uncertainties, on the one hand, and our incontrovertible verities on the other, choices become easy. All choices, regulatory, political, and ethical. Choices are much harder when we are less sure about all our facts, especially facts about the distant future. Believing as he does in markets and human ingenuity, the Hard Green is certain only that he is not too sure what tomorrow holds. Any undue sense of confidence conveyed in my earlier chapters I hereby retract. I am not prepared to chop hands on the strength of any of *my* facts. Of that I am quite sure.

---

* Karl Popper, *Conjectures and Refutations* (New York: Routledge, 1992), p. 35.

DESTINY

The gentle cow has journeyed with me the length of this book. Half nature, half man, she grazes unperturbed as Soft and Hard line up for battle at opposite ends of her pasture.

Not so long ago, she was an icon of wealth and security. If you were inclined to worship nature, you began with her. Today, Soft Greens view her as an environmental abomination. She is a wasteful engine for converting healthful plants to toxic meat and milk. She occupies the muddy pasture that was once a prairie or a beautiful wood. From both ends, her copious gut emits vast quantities of methane, a potent greenhouse gas. She symbolizes all the anti-green excesses of our Western appetites, McDonald's on the hoof.

Hard Greens don't doubt any of this. They merely doubt the cow will become extinct any time soon. Rich or poor, people are not going to stop craving milk and meat. So the Hard Green promotes the bionic cow, promotes efficiency in farming, as assiduously as the Soft promotes efficiency in refrigerators. Higher tech is greener.

But however bionic she may become, our cow will remain imperfectly contained by the market. She may still overgraze the common pasture, and she will still break wind copiously into the common air. Hard Greens understand market failure quite as well as Softs; T. R. himself learned some of his hardest lessons on his Dakota cattle ranch and the common range. Hard Greens part company with the Softs on how to set things right. Softs want to mandate and prescribe their way out of the Tragedy. Hards know externalities cannot be eliminated entirely, but are still committed to progressively moving things people value off the commons and onto private premises. At the same time, and without any great sense of self-contradiction, we can also promote the kind of intelligent, public conservation that T. R. pioneered. With high technology, private property, and public conservation we can have fewer cows, less Tragedy, and more meat and milk, too.

The Soft Greens say that more of everything is impossible, in the long run. If we run out of nothing else, we will run out of land. Thomas Malthus proved it. His was the original model of the future, sublimely simple and utterly compelling. It was beyond scientific, even: Malthus reduced the whole story to mathematics. Cow pasture increases arithmetically, human population geometrically, so we always run out of burgers in the end.

But only one thing fulfills a Malthusian prediction every time: a degenerate, big-future, big-government, political ideology. The predicted crisis in supply and demand inevitably materializes after bold government policies are put in place to avert it. The Marxists spread famine, poverty, and general misery from Angola to Kamchatka. In the environmental arena, they delivered half a planet of environmental ruin on a scale never approached by anyone else. For good measure, Stalin also promulgated coercive measures to increase population.

As Daniel Webster observed, there are men of every age who scheme to govern, who mean to govern well, but above all, who mean to govern, who promise to be good masters, but mean to be masters. Tellingly, perhaps, they are rarely good masters of themselves. Rousseau insulted, quarreled, and exploited his way through life, illiterate mistress in tow. After greeting the outbreak of the French Revolution with great enthusiasm, Condorcet ended up being hunted by Girondins. In hiding, he continued to write about the human race's uninterrupted advance toward enlightenment, virtue, and happiness. Two days after he was arrested he died in prison, whether from exhaustion or by poison is not known. Marx lived a fractious and parasitic personal life, manipulating, exploiting, and abusing everyone within his grasp, financially, sexually, and in every other possible manner. Some Soft Greens who have made cameo appearances in this book recycle their wives as readily as they recycle their glass bottles. Perfecting humanity is easier, it seems, than perfecting oneself.

Malthus knew better. He wasn't promoting utopian socialism, he was reacting against it. He dashed off his original pamphlet in response to French revolutionary theorists and British utopians, his own father among them. The full title of his 1798 pamphlet, which he published anonymously, was *An Essay on the Principle of Population as it affects the Future Improvement of Society, with Remarks on the Speculations of Mr. Godwin, M. Condorcet, and other Writers*. Those were the writers who confidently predicted, in Malthus's time, that rational men would soon live prosperously without the coercive institutions of family or private property. The thirty-two-year-old Malthus replied by noting the harsh social realities of his times and slapping down a two-line model that "proved" just the opposite.

As Darwin would later grasp, Malthus's model was roughly right, for almost every species but man himself. Far wiser than he is often made out to be, Malthus understood that, too: People have the capacity to choose. Epidemic, famine, and war will indeed maintain equilibrium of supply and demand, but only if wiser means don't. The "socialism" Malthus advocated was of the old-fashioned kind: morality, religion, internalized standards of good behavior. He favored free medical assistance and democratic institutions—radical ideas for his times—but urged that no welfare be administered in ways that would encourage decline in moral restraint. People should postpone procreation, he said, until economically able to support their children.

The more secure and affluent our own personal lives, the more we conserve and value life around us. The more we have, the more we are able to share, and the more willing we are to share, too. Traditional conservation, the green that does not depend on models and whose rewards are tangible today, not just a century from now: That kind of green is the environmental movement of the affluent. It is American Green, the green not of global environmental villains but of a confident, affluent people whose destiny it is to secure the peace among nations, and the green peace, too.

A
Conservative
Environmental
Manifesto

CONSERVATION IS THE POLITICAL HERITAGE OF THE CONSERVATIVE. We conserve paintings and manuscripts, furniture and buildings, churches and liturgies, social and ethical norms, political institutions and the rule of law. As a political movement, environmentalism was invented by a conservative Republican, Theodore Roosevelt. It was T. R. who so loved "silent places, unworn of man" that he became the first president wholeheartedly dedicated to conserving them.

We do not abandon conservative principle when we conserve wilderness, we affirm it. We affirm our own long tradition of creating parks, husbanding wildlife, and venerating natural heritages of every kind.

Scarcity

ECONOMIC SCARCITY

There is no inherent scarcity of food, fuel, metal, mineral, or space to bury our trash. We undoubtedly will exhaust some goods, some day. But we will grow, find, or invent other things to replace them. No law of geophysics, biology, engineering, or economics decrees: So far, but no farther. With ordinary economic goods, free markets and human ingenuity transcend all limits.

To avert economic scarcity, we will unleash markets.

GREEN SCARCITY

The one real and growing scarcity is scarcity of wilderness and the wildlife that dwells there, scarcity of forest, lake, and river, scarcity of marsh and shore, scarcity of places undeveloped by markets and untouched by the hand of man. What grows scarce isn't the land and livestock man can tame, cultivate, or exploit, but the land and life he chooses not to. The one true scarcity is not economic; it is un-economic.

We are committed to maintaining and extending uneconomic forests, uneconomic lakes, uneconomic shores, uneconomic wetlands, uneconomic wilderness.

Plenty

GOVERNMENT-PRESCRIBED EFFICIENCY

We reject the trickle-up theory of green, the vague notion that what we save in the more efficient refrigerator will somehow translate into savings at the power plant, which will translate into further savings at the mine-head. We do not believe that the way to conserve lakes, trees, mountains, and prairies is to conserve gasoline, newspaper pulp, aluminum, and glass, still less to sift and sort relentlessly through our own trash.

Government-prescribed efficiency programs are a distraction at best. They promote a false illusion of green progress. Efficiency imposed by government edict promotes nothing but inefficiency. It does not save wilderness.

We will leave the pursuit of efficiency to the market.

RECYCLING

Recycling is said to save land and raw materials. But perfectly good markets already set prices for those resources. If there is private profit to be found in modern trash, it will be found without government subsidy or compulsion.

The choice to recycle is best left to individuals and local communities. We believe, however, that most of the money and effort currently devoted to recycling would be better spent expanding publicly owned parks, forests, lakes, rivers, wetlands, and shores.

Hard Agriculture

A pest-resistant corn that doubles the farmer's yield per acre saves something that is environmentally important: land itself. So does a growth hormone that delivers more cow and milk on less pasture. Fertilizers, pesticides, packaging, and preservatives are the chemical keys to the best solar-power systems in widespread use today. They permit us to capture more energy from the sun, more efficiently, using less land. In return for the modest amounts of energy required to make them, they substantially boost the performance of the finest solar power engines on Earth: green plants. Hard agriculture transforms earth and sun, corn and wheat, chicken and cow, into edible calories much more efficiently than Soft alternatives. "Organic" food requires far more land to produce, and is therefore less green.

Preservatives and packaging reduce spoilage. Poor packaging that allows spoilage is as wasteful as a dripping faucet or a leaking gas tank. Irradiating chicken, meat, and spices kills salmonella and extends shelf life. Because most food is transported great distances, cutting spoilage does at least as much good for the environment as improving the gas mileage of the truck. More important, it effectively boosts the yield per acre of cultivated land.

We support Hard agriculture, because it is greener than Soft.

Hard Power

Hard power extracts more power from less of the Earth's surface. Uranium is harder than oil and gas, which are harder than coal, which is harder than biomass, solar, and wind. The greenest fuels are the ones that contain the most energy per acre of land covered, cultivated, paved, or stripped. Per unit of power produced, softer fuels consume more material, labor, and—above all, land. Policies that promote soft fuels over hard do not protect the environment; they hasten its destruction.

We support Hard power, because it is greener than Soft.

Hard Technology

Per unit of output, large centralized, industrial plants are usually much cleaner than the decentralized, low-tech alternatives they displace. It is far more efficient to burn oil in the huge, well-maintained boiler of a central power plant than to burn it in a two-stroke engine of a lawnmower, even after we allow for all additional losses in transmitting electricity from the power plant to the end user. It is more efficient, and cleaner, to burn fuel and distribute electricity than to refine fuel and distribute gas or gasoline.

The technology of atom and photon is the greenest of all. Nuclear power, the original "solar," extracts limitless energy from the tiniest amounts of material because it extracts subatomically. Irradiation is an excellent preservative because its energy disrupts the chemical bonds we want disrupted, the ones salmonella most needs to live. Improved genes in rice or a tobacco leaf substitute for a chemical factory producing phosphates or fertilizers, or a pharmaceutical plant manufacturing drugs. Hormones that promote growth and milk production in cows, pesticides that disrupt the hormonal systems of insects, are extremely efficient, and extremely safe, too, because they are so precisely aimed at such specific molecular targets.

We support Hard technology, because it is greener than Soft.

LIVING IN THREE DIMENSIONS

Hard technology saves the Earth because it enables us to live in three dimensions rather than two. We do not favor gathering more fuel from the surface, or traveling more on the surface, or spreading out unnecessarily over the surface, as Soft policies encourage. We promote the conservation of wilderness and the preservation of species by scraping less from the living surface and living more off the sterile depths and heights. We believe in digging up our energy, burying our wastes, flying high, tunneling deep, and thus leaving more of the surface undisturbed by man.

MARKET EFFICIENCY

Government "efficiency" edicts are never efficient. It is the free market that is efficient, spontaneously efficient. The Hard technology of modern capitalism is very efficient. In itself, Hard technology does not reduce consumption. It makes us richer, not poorer, and thus permits us to consume more. Efficiency—voluntary, by choice, impelled by market forces and free consumer choice—has only one definite, predictable effect: It boosts wealth. But wealth itself is green.

We support policies that promote economic wealth.

WEALTH IS GREEN

Wealth, not poverty, supplies the means to conserve wildlife, forest, seashore, and ocean. The charge that the rich are the despoilers, the exhausters, the expropriators of the planet's biological wealth is altogether false.

Poverty limits population, by letting people die. Wealth limits population by letting people live. Affluence allows parents to raise fewer, more robust children. Producing food abundantly is the most humane and effective way to limit population.

The rich man secures his genetic posterity through quality, not quantity. He puts his wealth into charity, art, bird feeder, and the prairies beyond. The only form of poverty that is green is the poverty chosen freely by the man so rich—so rich in capital, so rich in spirit—that he freely chooses to spend his treasure preserving and advancing more life than his own.

We affirm that wealth is greener than poverty.

Pollution

Much progress has been made with some of the worst sources of pollution: big smokestacks, car emissions, industrial effluents, sewage. But there is still much to be done.

Formulating effective abatement strategies is not easy. Simple-minded approaches often fail or backfire. Human pollutants are by-products of the market; they cannot be any more predictable or controllable, over the longer term, than the market itself.

TURNING POLLUTION INTO PROPERTY

Serious pollutants are best addressed in ways that neither expand the public sphere nor undermine established private rights. Issuing permits in quantities that mirror established patterns of activity and use is usually the best and most economically efficient way to proceed. Going forward, the market creates direct competitive substitutes for pollution permits—pollution abatement technology. The more freely people can buy, sell, and trade pollution and permits, the more pollution we will abate. Government and private parties can operate in such markets, too, buying up permits and retiring them.

We support the privatizing of pollution.

THE MICRO-ENVIRONMENT

Some micro-pollutants do real harm. When harms are reliably established, their sources should be contained.

But we reject all casual presumptions of guilt in the micro-environmental realm. Most of the time, micro-environmental pursuits are more wasteful than the waste itself. Most consume more energy, material, and time, endanger more lives, generate more pollution, and dissipate more value, than the things pursued. As a general rule, we are far better off putting our money and effort where clear environmental returns can be realized directly. Far more environmental good would have been achieved buying up green spaces, riverbank, watershed, and forest with the tens of billions that have been wasted instead on Superfund.

We will direct environmental spending away from the pursuit of micro-environmental phantoms and toward the conservation of parks, forests, lakes, rivers, wetlands, and shores.

NATURE'S ROLE

Spending our money on visible green objectives does more to advance invisible ones, too, because nature has such great power to cleanse, detoxify, and regenerate. Sufficiently dispersed in space and time, many pollutants are, in fact, harmlessly dissipated, broken down, and biologically recycled. The planet is vast, and nature is more robust than is sometimes imagined. Nature, in its fecundity, seems to have extraordinary power to recover and recapture landscapes leveled by man, or by nature itself, as a consequence of oil spill, forest fire, or volcanic eruption.

Wilderness conservation is generally the cheapest, most effective, and most pleasant way to cleanse the micro-environment. The most beautiful way to purify water is probably the most effective, too: Maintain unspoiled watersheds. The best way to suck carbon out the air is to grow trees.

Cleanup makes sense after an oil spill or other environmental accident. But cleanup is too often pushed to ruinous, anti-environmental excess. We will not indulge such excesses for the mere satisfaction of punishing polluters.

Politics

Takings

Markets are always cheating on their "little" pollutants; regulators are always cheating on their "little" takings. Markets fail; pollution is serious and real; and by all economic logic it should be abated. Government prescriptions fail, too; regulatory "takings" are serious and real; and by all constitutional logic, they must be abated.

When individual producers can dump the true costs of production onto the environment, they may create the illusion of prosperity as they impoverish society as a whole. When regulators can dump the true costs of government on the private sector, they may create the illusion of good government, even as they impoverish society as a whole. In markets and in government alike, returns don't just diminish, they go negative, when costs are palmed off surreptitiously on others, whether by smokestack or by sleight of the regulatory hand.

Conservative green begins with defining and permitting the status quo, in a way that does not redraw established lines between private activity and public interest. From that starting point, wilderness areas should be expanded, and pollution abated, on a pay-as-we-go basis, with markets in control.

Honest Bookkeeping

Government and the private sector alike must maintain honest green accounts. Environmental markets, markets for pollution and its abatement, require honest bookkeeping, just as other markets do. That means keeping track of credits as well as debits. Growing new trees, for example, removes carbon from the air. So does mummifying organic and plastic waste in landfills. If we are serious about our green objectives, we will keep honest books, with due allowances on both sides of the ledger. If we can't or won't, no real progress will be made, not through markets, and not by conventional regulatory prescription either.

We will both require and accept honest bookkeeping in all environmental matters. We will record debits and credits evenhandedly, where they are due.

Private Conservation

Private conservation is, by a wide margin, the most important form of conservation we have. Much of the time, effective conservation is possible on a scale that is commensurate with private ownership and control. We support private conservation initiatives wholeheartedly.

PUBLIC CONSERVATION

We recognize, however, that at some point the vastness of White Mountains and Everglades, of river archipelagos and coral reefs and the sheer scope and scale of the most ambitious conservation objectives require a reach to match. That means the reach of local, state, and federal governments.

We recognize that private fences cannot always conserve the value of the wilderness. Great, wide-open spaces are valuable because they are great and open. A vital part of Yellowstone's grandeur, and our own, is that it belongs not to Wall Street but to America. Value that inheres in citizenship, nation, patriotism: Such values cannot be contained or conserved in any private market. To privatize here is to destroy.

Government can play an essential role in husbanding and expanding the wilderness. The point of conservation is to be economically inefficient and unproductive, to retard conventional economic progress, not promote it, to do so in well-designated places, set aside for that specific objective. Conservative government can and should advance these objectives, where private ownership cannot.

SEPARATION OF POWERS

By embracing government's role in conserving uneconomic public goods, we grow even more adamant in opposing government meddling in ordinary economic affairs. It is because we want government to do its conservation well that we insist on shrinking government's role in agriculture, lumber, electric power, canals, railroads, highways, flood control, home mortgages, insurance, and more. Charge the government with promoting both wilderness and economic output, and the wilderness will be on the same road to ruin as the economy.

Government should promote uneconomic goods and leave economic ones to the market.

BIG GOVERNMENT

Taking control of the "environment"—literally "that which surrounds"—is an enormous political opportunity that the Left is eager to embrace. For the Left, it is an opportunity to be bureaucratic, manipulative, and pervasively intrusive, to take control of your lightbulb, flush toilet, and hair spray, your washing machine and refrigerator, your compost heaps and your cars. Nothing is too small, too personal, too close to home to drop beneath the Left's environmental radar. For people who like big government, this is political ambrosia.

But so long as people of that mind-set are in primary control, the environment will never be secure. Centrally planned industrialism, for all its desperate pursuit of efficient production, produced far fewer amps and ingots than honestly efficient capitalism. The people who couldn't deliver the electricity or the canned salmon won't deliver cool air or oceans teeming with fish, either.

Expanding the environmental bureaucracy does not protect the environment. We will shrink offices on the banks of the Potomac. We will expand protected forests, parks, lakes, wetlands, and shores.

ENVIRONMENTAL GEOPOLITICS

America can control pollution in its own rivers and local air, whatever happens in Brazil. We should save our cougars whether or not Brazil saves its jaguars.

But there is only one global climate, and there is, increasingly, only one global market for steel, petrochemicals, aluminum, and many other energy-intensive, air-devouring, water-fouling industries. It is no use fencing in just the north side of a Commons, if it is just a short walk around to the unfenced south. Relocating steel factories to China and petrochemicals to Dubai will not help the environment; it is far more likely to harm it. Overfishing of the oceans—a serious problem—will not be solved by substituting Japanese dragnets for American ones.

It is a fraud to set about solving a "commons" problem piecemeal, here but not there, north but not south, west but not east, one industry but not another, debits but not credits. The commons problem arises *because* a resource is shared, so that no one acting alone has incentive or ability to take care of it properly. Solutions that aren't equally uniform, equally "common," are a complete waste of time. Half a fence solves nothing; it just forces law-abiding people to take a hike, or sell their herds to their lawless neighbors.

We support transforming environmental resources into property rights under the jurisdiction of existing national government authorities: responsible, democratic governments that affirm the rule of the law. We do not believe in negotiating the environmental future with the many thieves, liars, and bandits that call themselves statesmen and crowd the halls of the United Nations. The Fidel Castros of the world cannot be trusted to affirm property, regulatory norms, or anything at all remotely related to the rule of law. Environmental interests will be set back, not advanced, by confiding any authority to such people.

ETHICS

We are not pagan worshippers of Mother Earth. We accept the traditional Judeo-Christian teaching, that man and nature are not equal. Our interests in nature are aesthetic, not moral. Our moral imperative is to put people first whenever direct choices have to be made.

But aesthetic standards can and must be maintained, too, even in an imperfect world in which humans still suffer. We may build cathedrals and preserve great art, even in times when some are still homeless, and so we should. We may conserve nature, too, and so we should.

It is good to conserve what is beautiful in man's own creation. It is good to conserve the boundless wonder and beauty of God's.

ACKNOWLEDGMENTS

This book was written under the auspices of the Civil Justice Program of the Manhattan Institute for Policy Research. I am deeply grateful to the Institute's president, Larry Mone, for his patient, unstinting support and help.

I first developed some of the thoughts and text of this book in articles and columns that appeared in *Regulation*, *Columbia Law Review*, *Virginia Law Review*, *Harvard Law Review*, *Forbes*, and *Commentary*. I owe much to those journals' editors for their help along the way, most particularly to Bill Baldwin of *Forbes* and Neal Kozodoy of *Commentary*. For *Hard Green* itself, Elizabeth Anderson provided excellent research assistance and editing all along the way. My thanks, also, to Shad Gohn and Marilyn Leeland, who helped with the final stages of the editorial process.

NOTES

PREFACE

1. David Gelernter, "The Immorality of Environmentalism," *City Journal* (Autumn 1996): 24.

2. All my references to Brands are drawn from H. W. Brands, *T. R.: The Last Romantic* (New York: Basic Books, 1997), 11, 186–188, 207, 623–626, 646–648.

3. Daniel Webster, quoted in Nominations of Abe Fortas and Homer Thornberry (Nomination of Abe Fortas to be Chief Justice of the United States), Hearings Before the Committee on the Judiciary, United States Senate, 90th Congress, 2nd Session, 7/11–7/23/68, 108 (Washington, D.C.: U.S. Government Printing Office, 1968).

CHAPTER 1

1. All my references to Ehrlich are drawn from Paul Ehrlich, *The Population Bomb* (New York: Ballantine Books, 1968) xi; and Paul Ehrlich and Anne Ehrlich, *Betrayal of Science and Reason: How Anti-Environmental Rhetoric Threatens Our Future* (Washington, D.C.: Island Press, 1996), 73–74, 101–103.

2. All my references to Meadows are drawn from Dennis Meadows and Donella Meadows, *The Limits to Growth* (New York: Universe Books, 1972), 193; and Donella H. Meadows, "Charting the Way the World Works: Planning a Global Computer Model," *Technology Review* (February 1985): 54.

3. All my references to Cairncross are drawn from Frances Cairncross, *Costing the Earth: The Challenge for Governments, The Opportunities for Business* (Boston: Harvard Business School Press, 1992).

CHAPTER 2

1. All my references to Hardin are drawn from Garrett Hardin, "The Tragedy of the Commons," *Science* 162 (1968): 1243; and Garrett Hardin, "Living on a Lifeboat," *BioScience*, (October 1974): 561–568.

2. All my references to Gore are drawn from Albert Gore, *Earth in the Balance: Ecology and the Human Spirit* (Boston: Houghton-Mifflin, 1992), 39–40, 127–128, 183, 362–363.

3. All my references to Lovelock are drawn from James Lovelock, *Gaia: A New Look at Life on Earth* (New York: Oxford University Press, 1995), x, xiv, 21, 26, 31, 75–77, 110, 119.

CHAPTER 3

1. All my references to Lovins are drawn from Amory Lovins, "Energy Strategy: The Road Not Taken," *Foreign Affairs* (October 1976): 65–96; Amory Lovins and L. Hunter Lovins, *Brittle Power: Energy Strategy for National Security* (Andover, Mass.: Brick House Publishing, 1982), 1, 19; Amory B. Lovins and L. Hunter Lovins, "REAL Security," *Daughters of the American Revolution Magazine* (March 1983): 242–245; and Brad Lemley, "Power to the People: Amory Lovins sees the Big Energy Picture—And the Secret, He Says, Is to Think Small," *Chicago Tribune*, August 3, 1998, C9.

2. All my references to Perrow are drawn from Charles Perrow, *Normal Accidents: Living with High-Risk Technologies* (New York: Basic Books, 1984), 11, 146, 173–174, 230, 348, 351.

3. Herbert A. Simon, "On a Class of Skew Distribution Functions," *Biometrika* 52 (1955): 425–440.

CHAPTER 4

1. Thomas T. Samaras, *The Truth About Your Height: Exploring the Myths and Realities of Human Size and Its Effects on Performance, Health, Pollution, and Survival* (San Diego: Tecolote Publications, 1994).

2. Malcolm W. Browne, "Unearthing an Avalanche of Ideas That Rose from Energy Crises Past," *The New York Times*, April 14, 1998, F4.

3. Richard Klein, *Eat Fat* (New York: Pantheon Books, 1996).

CHAPTER 5

1. James Watson and Francis Crick, "Molecular Structure of Nucleic Acids: A Structure for Deoxyribose Nucleic Acid," *Nature* 171 (April 2, 1953): 737.

2. Richard Dawkins, *The Blind Watchmaker* (New York: Norton, 1986).

3. Stephen Jay Gould, *Full House: The Spread of Excellence from Plato to Darwin* (New York: Harmony Books, 1996).

CHAPTER 6

1. Alston Chase, *Playing God in Yellowstone: The Destruction of America's First National Park* (Boston: Atlantic Monthly Press, 1986).

2. James Brooke, "Land Trusts Growing Faster in Rockies and Southwest, Study Shows," *The New York Times*, October 1, 1998, A18.

CHAPTER 8

1. Richard A. Houghton and George M. Woodwell, "Global Climatic Change," *Scientific American* (April 1989): 36, 38.

CHAPTER 9

1. Murray Feshbach and Alfred Friendly, Jr., *Ecocide in the USSR: Health and Nature Under Siege* (New York: Basic Books, 1992): 88.

CHAPTER 10

1. Dana Milbank, "In His Solitude, a Finnish Thinker Posits Cataclysms," *The Wall Street Journal*, May 20, 1994, A1.

2. James Taranto, "Why the Unabomber Must Die," *The Wall Street Journal*, January 6, 1998, A18.

3. Paul Johnson, "The Perils of Risk Avoidance," *Regulation* (May/June 1980): 15–19.

CHAPTER 11

1. Peter Schwartz and Peter Leyden, "The Long Boom," *Wired* 5.07 (July 1997): 116.

2. Isaac Deutscher, *Stalin: A Political Biography*, 2d ed. (New York: Oxford University Press, 1967).

INDEX